SpringerBriefs in Geography

SpringerBriefs in Geography presents concise summaries of cutting-edge research and practical applications across the fields of physical, environmental and human geography. It publishes compact refereed monographs under the editorial supervision of an international advisory board with the aim to publish 8 to 12 weeks after acceptance. Volumes are compact, 50 to 125 pages, with a clear focus. The series covers a range of content from professional to academic such as: timely reports of state-of-the art analytical techniques, bridges between new research results, snapshots of hot and/or emerging topics, elaborated thesis, literature reviews, and in-depth case studies.

The scope of the series spans the entire field of geography, with a view to significantly advance research. The character of the series is international and multidisciplinary and will include research areas such as: GIS/cartography, remote sensing, geographical education, geospatial analysis, techniques and modeling, landscape/regional and urban planning, economic geography, housing and the built environment, and quantitative geography. Volumes in this series may analyze past, present and/or future trends, as well as their determinants and consequences. Both solicited and unsolicited manuscripts are considered for publication in this series.

SpringerBriefs in Geography will be of interest to a wide range of individuals with interests in physical, environmental and human geography as well as for researchers from allied disciplines.

More information about this series at http://www.springer.com/series/10050

Tarje I. Wanvik

Contested Energy Spaces

Disassembling Energyscapes of the Canadian
North

 Springer

Tarje I. Wanvik
Department of Geography
University of Bergen
Bergen, Norway

ISSN 2211-4165   ISSN 2211-4173   (electronic)
SpringerBriefs in Geography
ISBN 978-3-030-02395-9   ISBN 978-3-030-02396-6   (eBook)
https://doi.org/10.1007/978-3-030-02396-6

Library of Congress Control Number: 2018960446

This Springer imprint is published by the registered company Springer Nature Switzerland AG
The registered company address is: Gewerbestrasse 11, 6330 Cham, Switzerland

Preface

For decades, extractive industry developments have had direct and indirect impacts on indigenous communities in Wood Buffalo, AB, Canada. Yet, in a seemingly paradoxical manner and despite massive negative attention, there are several indigenous communities in favour of industrial developments on their traditional lands. This book constitutes an exploration of the contested energy space of the Canadian oil sands—analysing the characteristics, governance and power plays therein. I investigate how to *conceptualize* the *socio-material complexity* of contested energy spaces in the Canadian north, to identify *instability* and *potential for change* within them and to *understand the power relations* between industries, state and indigenous communities. Hence, the overall effort of this book transcends the apparently narrow issue of indigenous responses to industrial impact, touching upon larger, more complex and generic problems of energy and society relations.

Bergen, Norway   Tarje I. Wanvik

Acknowledgements

I have indebted myself to many good people during the 4 years of planning, researching and writing this book. First and foremost, an extended thank you goes out to my three supervisors: Knut Hidle, Håvard Haarstad and Jørgen Ole Bærenholdt. They have all been contributors of great inspiration, thorough correction and knowledgeable guidance.

However, without some invaluable assistance in the field, this effort would have been nothing more than a stranded ambition. I am forever grateful to Peter Fortna, Jay Telegdi and Lena Gross for their hospitality and kind reception of a lost, rookie Norwegian researcher on the prairie. They invited me into their homes, shared their experiences and introduced me to some fantastic, friendly and knowledgeable people among the Métis communities of Fort McMurray, Fort McKay and Conklin. Special thanks go to Bill Loutitt, Jeffrey O'Donnell, Ron Quintal, Dan Stuckless and Dwayne Trudeau Roth.

Special appreciation also goes to Statoil Canada (now Equinor) for letting me in on some of their corporate social responsibility activities and directing me to their piles of thorough assessments and reports on environmental impacts.

Fortunately, I have been blessed with a crowd of admirable people around me, inspiring me, cheering me on and being tremendous role models for this academic pursuit: Øyvind Paasche, Simon Pahle, Camilla Houeland and Stian Suppersberger Hamre. I would also like to thank the fabulous research team at Spacelab, Stina Oseland, Kristin Kjærås, Jakob Grandin, Karin Lillevold, Siddharth Sareen, Hanna Kvamsås and Stefan Bouzarovski, and my good colleagues and friends in the Department of Geography. Sincere thanks also go to the administrative staff for pertinent support and to all my students over the past 4 years, who have challenged me and given me high hopes for the future of geography.

This research would not have been possible without core funding from the IMER research network and generous travel support from the Meltzer Foundation.

Finally, I am forever grateful to my family: to my father and mother for reading and commenting along the way; to my wife and life companion, Torill, for reigniting my academic flame; and to our amazing daughters, Laura and Anna, for keeping up with me and inspiring me every single day. I love you all.

Contents

Chapter 1
Introduction: Contested Energy Spaces

Abstract For decades, extractive industry developments have had direct and indirect impacts on indigenous communities in Wood Buffalo, Alberta, Canada. Yet, in a seemingly paradoxical manner and despite massive negative attention, there are several indigenous communities in favour of industrial developments on their traditional lands. To investigate this paradox, the author embarks on an exploration of the contested energy space of the Canadian oil sands—investigating and analysing the characteristics, governance and power plays therein.

Keywords Energy · Extraction · Canada · Indigenous · Instability · Governance

In this book, I have investigated how to conceptualize the socio-material complexity of contested energy spaces in the Canadian North, how to identify instability and potential for change within them, how to understand the power relations between industry, state and indigenous communities and finally how to conceptualize non-human agency within these complex landscapes of energy production. All with the ambition to understand and explain the complexity of contested energy spaces.

By employing assemblage thinking, I identify contested energy spaces as complex places or situations. I argue that to analyse and understand these complex situations, we need to equip assemblage theory with acknowledged geographical concepts of place (and materiality), scale (and networks) and power (as the mobilization of resources), providing analytical categories and tools for geographers investigating contested energy spaces specifically, and hopefully also contributing to the ongoing scholarly discourse on place.

Furthermore, I investigate how to identify *instability* and *potential for change* in contested energy spaces. I elaborate on the instabilities of contested energy spaces, underscoring that instead of talking about techno-institutional complexes, regimes or a coherent systemic "fossil capitalism" held together by co-articulation of institutions, infrastructures and practices, we can talk about a looser association of different social and material elements drawn together and pulled apart by a range of different forces. I argue that this is liberating because it frees us from the assumption that changes need to have an impact on the fundamentals of larger socio-technical

T. I. Wanvik, *Contested Energy Spaces*, SpringerBriefs in Geography,
https://doi.org/10.1007/978-3-030-02396-6_1

regimes to be significant. For me, the important point is to illustrate that contested energy spaces are fragmented, contested and converted at particular sites. Therefore, contradicting those who suggest that assemblage thinking blunts critical sensibilities, I find that it is helpful in opening spaces for negotiation and contestation. I argue that there is a normative rationale for shifting researchers' attention towards instability and change. Destabilizing the permanence of contested energy spaces may be productive in its own right. The emphasis on structural constraints runs the risk of reproducing the oil industry's carefully scripted narrative of its own inevitability. It is critical that the specific lens that spatiality affords geographers is also used to identify the cracks in the wall and the leverage points for transformation.

I proceed through three case studies to understand the power play between industry, the state and Indigenous communities in the contested energy spaces of the Canadian North, but from different perspectives, or on different scales. In the first case, I show that industrial activities have had great impacts on the social, cultural and environmental realities of the contested energy spaces. The burden has been substantial for local communities and has added to the prolonged historical conflict between the Crown and Indigenous communities over rights and entitlements. This complex relationship has led to substantial challenges for all stakeholders. In response to these challenges, the federal duty to consult, along with provincial environmental impact assessments (EIA) and locally negotiated impact benefits agreements (IBA), has been delegated to industry, where corporate social responsibility (CSR) and stakeholder management form important centrepieces. This delegation has been legitimized on pragmatic grounds to underscore the better positioning of industry to consult Indigenous communities, to assess its own impact and to negotiate compensation and benefits agreements. I identify an interrelated, nested and multiscalar governance structure emerging from these four distinct governance features (consultations, EIA, IBA, and CSR) that can be viewed as a joint mobilization effort by government, extractive industry proponents and indigenous communities to realize a workable, win–win regulatory environment in the contested energy space of Wood Buffalo.

In the second case, Indigenous communities calibrate their participation in the emerging governance processes in the contested energy space of Wood Buffalo to strengthen their negotiating power. I take assemblage thinking as the basis of an analytical framework to examine Indigenous Métis communities in Wood Buffalo. I reveal that Indigenous engagement with extractive industry development is neither static nor (only) responsive in character. Rather, Indigenous communities are strategic pragmatists that creatively and proactively engage in the development of extractive industries in their traditional territories. Viewing the interactions between the component parts of the contested energy space of Wood Buffalo as the workings of an unstable and changeable assemblage reconfigures our interpretation of Indigenous engagement; we no longer see the people as passive victims or as only responsive to external pressure; we now see Indigenous communities as proactive, pragmatic component parts of the Wood Buffalo energyscapes. I show that through strategic pragmatism, their traditional ways of life are imbued with substantial transformative capabilities. I show that these capabilities have moved the Métis communities of Wood Buffalo into formalized alliances with other stakeholders striving to evolve and change, to harvest strategic resources to their benefit.

The Indigenous communities of this study to some extent favour high-impact industrial activities in their traditional territories for several specific reasons. First, the complexity exposed in contested energy spaces does not offer simplistic or conventional understandings of Indigenous agency. Second, the governance innovations of the contested energy space of Wood Buffalo entail different and untraditional approaches by which different stakeholders seek benefits from a highly lucrative industrial adventure. Third, by underscoring the instability of contested energy spaces and their constituent parts, I show that Indigenous communities are no less adaptable or pragmatic than other stakeholders, and they strive to evolve and change to harvest strategic resources for their betterment.

Finally, I investigate the role of non-human agency in contested energy spaces. As geographers, we seek to understand the intimate relationship between people and places, and this is equally valid and relevant within contested energy spaces. By following the aftermath of the Fort McMurray wildfire in 2016, I analyse how the fire itself led to substantial changes within the organized mobilization of resources among Indigenous communities of Wood Buffalo.

1.1   Disassembling Contested Energy Spaces

Energy spaces—sites where energy is excavated, harvested, produced or consumed—have been subject to human imaginaries since time immemorial. Because it is a product of resources and forces of nature, energy is inextricably linked to our relationship with our environment: to identity, belonging, ways of life and leisure. From the prehistoric fireplace to modern nuclear power stations, from the deep coal mines to solar plants and high wind turbines, energy has been a prerequisite for our very existence.

Perhaps we should not then be too surprised by the persistent contestation and debate around questions of energy spaces. Large geographic spaces have been expropriated and transformed into power engines for entire societies, changing land features and land use forever. The stories seem the same all over the contemporary world: almighty industrial energy complexes looking for new energy-producing opportunities are challenged by small local communities affected by industrial violations to their everyday lives and environments. The affected communities reach the headlines of media providers; alliances are forged, protests are held, statements are signed, but in the end, industry wins. The prevailing narrative of the contested energy spaces describes an uneven, skewed power relationship between powerful industries and nation-states on the one hand and powerless communities and people on the other.

This is particularly true for the Canadian North, which is the focal point of this research. In the mass media and academic literature, the contested energy spaces of

Canadian bitumen extraction[1] have been portrayed as chaotic and disorderly; they are both literally and metaphorically built on sand; tensions and a series of disputes over land and rights have arisen between the state, industry and local indigenous communities. Canadian governments have long exploited the bountiful natural resources of the land while at the same time attempting to reconcile a difficult relationship with local Indigenous communities. The oil sands region of Canada is primarily situated in the north-eastern part of the province of Alberta, until recently relatively sparsely populated by various Indigenous groups of First Nations or Métis origin.

By exposing its hostile features, vast tailing ponds and huge open wounds in the boreal forest landscape to the world (Szeman, 2012), the local and global media have made resource depletion and environmental destruction the primary imaginary of the oil sands. For decades, these developments have had direct and indirect impacts on indigenous communities throughout the region. However, contrary to this negative coverage, my initial observations and pre-field-research findings were inconsistent with the majority of existing literature on these energy spaces. In a seemingly paradoxical manner, I found that several Indigenous communities spoke in favour (albeit hesitantly and conditionally) of these industrial developments in their traditional lands. This sparked my curiosity, and I began developing my research questions and searching for answers. I embarked on an exploration of the contested energy space of Wood Buffalo, Canada—investigating and analysing its characteristics, governance and power plays to investigate this paradox (Fig. 1.1).

1.2   Research Questions

Contested energy spaces are fundamentally geographical. For centuries, geographers have described and analysed energy landscapes. The actors operating these spaces span a wide spectrum of geographical scales, from the local community to the global industrial conglomerate exploiting the resources. Therefore, geography is particularly suitable as a cradle for new, empirically grounded theoretical innovations around contested energy spaces.

My thematic focus is a contested field of research, involving a multitude of interpretations and meanings assigned to different actors and agitators. I wanted to look into the background of the energy production in the highly contested energy space of Wood Buffalo, Alberta and to consider its complexities, the historical and contemporary relationship between industry, state and indigenous communities, and how power plays manifest in the landscape. The contested energy spaces stretch far beyond the geographical location where extraction takes place. This implies that the *local* quickly emerges as a particular set of relationships and networks that link to a

---

[1] Proponents and opponents label this contested energy space differently. While proponents underscore the refined end product (oil), the opponents focus on the polluting properties of the source (tar sands).

Fig. 1.1 Field area (map design by Max Koller). Source: Natural Earth Data and Alberta Environments and Parks

much broader set of relationships rather than as a singularity focused on a bounded location. The aim of this book has been to conceptualize the socio-material complexity of contested energy spaces, to understand the power play between industry, state and indigenous communities, and to try to identify instability and potential for change within these landscapes. Ultimately, I have wanted to explore why some indigenous communities support extractive industry developments on their traditional territories, despite substantial destruction of the local environment and traditional indigenous land use practices. This research has brought me to head offices in Calgary and Edmonton, to corporate headquarters in Oslo, and all the way to the trap lines of Métis elders.

Valuable insights may be drawn from this research. Today, there is a growing demand for a more sustainable energy future. We need to rethink and transform the ways that we search for, produce and consume energy. However, to change, we need to scrutinize and understand our current contested energy spaces properly. Where are the weaknesses, where are the strengths and where are the pitfalls or the opportunities in the system when we opt for transformation and change? It is paramount to develop a theoretical and analytical framework to analyse barriers and challenges related to energy production and transformation. Existing contested energy spaces constitute excellent fields of study for such theoretical experimentation and production.

References

Szeman, I. (2012). Crude aesthetics: The politics of oil documentaries. *Journal of American Studies, 46*(2), 423–439.

Chapter 2
Understanding Contested Energy Spaces

Abstract Energy spaces have always been contested. A series of bitter disputes between extractive companies, host governments and local communities in recent years have led to heavy losses for investors and threatened the development prospects of many resource-rich countries. The management of tensions and risks around resource extraction is more or less the modus operandi of the extractive industries, just as fierce opposition to them seems to be a modus vivendi for local communities and NGOs all over the globe. In this chapter, the author develops an understanding of the complex characteristics of contested energy spaces as scaled and assembled; being both bountiful emptiness and cultural spectacle.

Keywords Energy · Extraction · Canada · Indigenous · Stakeholder management · Scale · Assemblage

Energy spaces, where primary resources such as oil and gas are extracted, are often objects in the global power play for energy deposits, and arenas of everyday life for those who live there. After decades of increasing energy prices, these energy spaces have attracted a wave of new investments in natural resources, especially in coal, oil and gas, and there are high expectations about the profits from the sector in many parts of the world. In the post-industrial globalization discourse, many have pointed out that these energyscapes are constructed as asocial voids, ghost sites or deserts (Bridge, 2004; Catton, 1982; Ferguson, 2005; Hetherington, 1997; McClintock, 1995). There is a certain "discursive cleansing" that we may recognize from the colonization of America. The sites are attributed to the emptiness of human beings or culture while at the same time providing value through fertility and resource overflow. It is on the basis of this understanding of energy spaces that Bridge (2001) coins the phrase "bountiful emptiness".

At the same time, theory concerning energy spaces has become more complex. The strong media and civil society focus on conflicts between multinational companies and communities has raised awareness about the local effects of natural resource extraction (Stevens, Kooroshy, Lahn, & Lee, 2013). The role of multinational companies in trade in controversial natural resources, tax evasion and environmental degradation in developing countries has been the theme of television documentaries, social media and established news sources. In the media, the "tar sands" in Canada, farmed salmon in Chile and diamonds in Sierra Leone, among many others, have reached the headlines. This has made it clear that the idea of "bountiful emptiness" cannot be maintained—such places are not asocial "locations" abounding with natural resources but otherwise empty. The energy spaces are instead understood to be arenas filled with social materiality, life and sensitive ecosystems.

These performances are still largely discursive—a range of stakeholders design these spaces in accordance with their interests, often via mass media communication. Indigenous peoples are particularly vulnerable to such media representation as "ecologically noble savages" living in covenant with nature and the places where they live (Redford, 1991). Logan and McNeish (2012) claims that this has created incentives for communities in conflict with energy companies to transform their demands and identities in line with stereotypes that arouse sympathy and attract support from non-governmental organizations (NGOs). The effect of this is to emphasize the cultural diversity of people who populate the energy spaces. The energy spaces thus become places where unique cultural traditions are expressed. Such an understanding of an energy space may be called a "cultural spectacle".

The extreme ways to understand energy spaces thus become "bountiful emptiness" on the one hand and "cultural spectacles" on the other. Simply put, it may be said that a bountiful emptiness comes from a global perspective or a bird's-eye view, where the places gain importance through their position in the worldwide energy industry. A cultural spectacle, on the other hand, is created on a local scale, as a grounded perspective on cultural differences. Clearly, both of these understandings have aspects that are important in understanding contested energy spaces. At the same time, they are simplistic, stereotyping and—not least—static, so they are not suitable for academic understanding or conceptualization (Fig. 2.1).

The dichotomy between the "bountiful emptiness" and "cultural spectacle" views derived from the global and the local scales, respectively, provides challenges in many arenas. Politically, it challenges the practices of extraction concessions, reciprocity and the power plays between companies and local communities and authorities. Academically, it challenges methods of data collection and the framing of the relevant situations (see Chap. 4 for details). Hence, to understand the scale of contested energy spaces, we need to understand the debates surrounding geographical scale in recent years. In the following section, I explain how geographical *scale* influences the notion of contested energy spaces.

2.1 Contested Energy Spaces as Scaled

What is scale, and how does it relate to contested energy spaces? In the most conventional understanding, scale is defined simply as the spatial reach of actions. Actions on different scales have different patterns, logics and rationalities, and

Fig. 2.1 Illustrating bountiful emptiness versus cultural spectacle. Above: Syncrude oil sands. Below: Métis trap line cabin. Photos by the author

deploy different material mediums and discursive idioms (Xiang, 2013). This is why, contested energy spaces have such different meanings when viewed on a global scale (bountiful emptiness) versus a local scale (cultural spectacles). Two types of scales are particularly important in contemporary social science: taxonomical and emergent. Taxonomical scales are the building blocks of "the nested hierarchy of bounded spaces of differing size, such as the local, regional, national, and global" (Delaney & Leitner, 1997, p. 93). However, life is of course more complex than taxonomy. In social life, in particular, places always reach far beyond their perceived boundaries. This brings us to the second type of scale—the emergent. An emergent scale is the scope of co-ordination and mobilization that arises from collective actions, which in turn generate new capacity for the actors. As an explicit theoretical project, constructionist perspectives on scale are a fairly recent development of geographical thought. Their emergence may be traced to broader changes in social theory, such as different understandings of power and practice, and wider acceptance of some version of "social constructivism", and as a response to the inadequacy of inherited conceptions of geographic scale for understanding profound and perplexing transformations in the contemporary world (Delaney & Leitner, 1997).

Building on the assertion that scale is better understood dialectically than hierarchically, I argue that in addition to aspects of size and level, geographical scale should be considered as having an important facet of relation. By considering aspects of scale such as relation, we may begin to fill some of the gaps left by too narrow a focus on size and level as the metaphorical facets of scale (Richard Howitt, 1998).

The scale debate, ignited by Marston et al. and their controversial article "Human geography without scale" (Marston et al., 2005), has preoccupied the geographic community for the past decade. They claimed that any a priori conceptualization of scales (or any other form of hierarchical socio-spatial formation) in human geography was at best simplistic; worse still, it was structuralist—a reification of a vertical power structure rendering local places and minorities powerless in the face of forces on higher scales (i.e. global). Instead, they advocated for a "flat ontology", a theory of assemblages and of heterogeneous, irreducible complexity. The language itself was indicative of this aim: flat versus hierarchical, horizontality versus verticality, self-organization versus structuration, emergence versus transcendence, attention to ontology as opposed to epistemology and so forth (Escobar, 2007).

Manuel DeLanda developed the framework of assemblage theory, which combines heterogeneous entities into some form of temporary relation (or set of relations), without presupposing that these relations necessarily constitute an organism (Anderson, Kearnes, McFarlane, & Swanton, 2012; DeLanda, 2006; Deleuze & Parnet, 2007). This theory evolved from two of the most important sources of flat ontologies: theories of complexity, particularly in the natural sciences, and the philosophy of Deleuze and Guattari. Flat alternatives can also be seen as building on and responding to the various waves of social constructionism and discursive approaches of the past few decades; they are akin to neo-realist ontology (Anderson et al., 2012).

The scalar protagonists have strongly confronted Marston et al., accusing them of selective reading of the scale literature, combined with a selective amnesia concerning the previous decades of scholarly debate. Particularly interesting in regard to this book are the arguments posed by Escobar (2007). Although an advocate of flat ontologies, Escobar argues that DeLanda never rejected scales as socio-spatial formations of some sort. Conventional approaches assume the existence of two levels (micro, and macro) or a nested series of levels (like a Russian doll). The alternative approach posed by DeLanda has been to use bottom-up analysis to show on each scale how the properties of the whole emerge from the interactions between parts, bearing in mind that the simpler entities are themselves assemblages of sorts.

Another highly relevant claim is made by Leitner and Miller (2007), who suggest that the recognition of scalar orders and existing power asymmetries is crucial to progressive politics, in terms of both the development of alternative political spaces and the deployment of socio-spatial strategies of resistance.

Like previous advocates of a scalar perspective, Jessop et al. (2008) question any form of privileging a single dimension of socio-spatial relations, scalar or otherwise. They believe that this contributes to (…) short intellectual product life cycles for key socio-spatial concepts, limiting opportunities for learning through theoretical debate, empirical analysis and critical evaluation of such concepts. They point to four distinct spatial lexicons developed by social scientists over the past 30 years: territory, place, scale and network (Dicken, Kelly, Olds, & Yeung, 2001; Paasi, 2004; Sheppard, 2002). No lexicon on its own can fully describe socio-spatial events (Jessop et al., 2008). This critique particularly targets the flat ontologies of assemblage theory, with its bias towards network centrism, one-sided focus on horizontal, rhizomatic, topological and transversal interconnections of networks, frictionless spaces of flows and accelerating mobilities (Castells, 1996; Sheller & Urry, 2006).

I believe that there is a middle ground here that is not properly addressed owing to disciplinary quibbles. I would advocate an empirically driven, multidimensional assemblage approach where the processes constituting an assemblage and the forms of organization of heterogeneous entities (situated, scaled and networked) constitute the analytical basis of any geographical analysis of particular empirical situations. In the next section, I elaborate on the notion of contested energy spaces as assemblages.

2.2   Contested Energy Spaces as Assembled

In recent years, the notion of assemblage has been explicated by Michel Callon when describing (economic) agency (Callon, 2008). Originally, Deleuze and Guattari employed the French term *agencement* to describe their ideas. *Agencement* in French translates as "arrangement", or what has come to be known as "assemblage". It conveys the idea of a combination of heterogeneous elements that have been carefully adjusted to one another. Assemblages are arrangements endowed with the capacity to act in different ways depending on their configuration (Palmås, 2007). Heterogeneous, multidimensional assemblages are loose gatherings of human and non-human entities interacting in ephemeral and emerging constellations, where the very process of interaction gives rise to new reactions, not only from within the assemblage but also from outside. A place can be an assemblage. A company can be an assemblage. Moreover, according to Manuel DeLanda, there are assemblages within assemblages (DeLanda, 2006, 2016), so they may extend outwards like networks or up and down a scale like hierarchies.

An assemblage, according to Deleuze, is a multiplicity made up of many heterogeneous parts that establishes liaisons and relations across ages, sexes and reigns— between constituent parts of different natures. Thus, the assemblage's only point of unity is that of co-functioning; it is a symbiosis (Deleuze & Parnet, 2007, p. 52). For

Deleuze, the "unity" of assemblage is not that of an organic whole, or of a total system, where different parts are smoothly or violently subsumed into homogeneity. Therefore, it can be distinguished from models of socio-political composition that draw on organismic metaphors (Anderson et al., 2012).

Component parts are involved in processes of territorialization and de-territorialization of the assemblage. Territorialization is a process that stabilizes the identity of the assemblage, whereas de-territorialization destabilizes it. This corresponds to traditional understandings of the concept of territory as the bordering, bounding and enclosure (Jessop et al., 2008) of an assemblage. According to Deleuze and Guattari, assemblages are entities that consist of bodies and objects (referred to as content), as well as non-material entities such as statements (referred to as expression). Assemblages can thus be characterized by ongoing processes of territorialization and de-territorialization. There are processes that stabilize/consolidate and destabilize/dissolve (respectively) the identity of the assemblage (Deleuze & Guattari, 1988).

Contested energy spaces could be viewed as assemblages where various entities across scales (companies, NGOs, civil society organizations, opinion leaders, indigenous groups, local government or national government) fight over territory (bountiful emptiness versus cultural spectacle), resources (oil, gas, water, cheap labour versus hunting, fishing areas, open spaces or good jobs) and political priorities (extraction versus protection). The encounters between assemblages are critical phases in these processes, so events or moments where assemblages meet can be saturated with tensions and friction (Tsing, 2005), and processes of territorialization and de-territorialization are most frequent. These encounters, or relations, are external to their terms, meaning that something new is created out of these connections and is contested and fought over by the different parties involved in the encounter. Contested energy spaces are examples of venues for such encounters and tensions.

Assemblage theory can be described as schizophrenic, inasmuch as it favours process over entities while insisting on the autonomy of parts and the exteriority of relations (Anderson et al., 2012). However, in reality, it invites a dissection of power relations and influences.[1] To understand how power and influence flow between the players in a given place or event, it is necessary to scrutinize the processes and to identify the transforming and/or stabilizing processes within and between assemblages over space and time. To conduct this power analysis, it is necessary to identify the networks, to explore the social hierarchies and to examine the local context. The next section expands on the notion of movers and shakers of energy spaces.

---

[1] Several other theoretical frameworks have similar objectives, such as Bruno Latour's Actor-Network theory, different varieties of socio-technical system theory and hybridity literature. However, I find DeLanda's reading of Deleuze and Guattari more attuned to the instability of relations: how seemingly powerful alliances are fragile and subject to change (for an extended discussion, see Haarstad & Wanvik, 2016). This has been important to my project of challenging established truths within resource governance in Alberta. At the same time, he is preoccupied with what he calls the "emergent capacities" of associations or what he terms encounters. This has been important to my introduction of pragmatism as an analytical concept to try to understand indigenous agency.

2.3 The Movers and Shakers in Assemblages: Stakeholders

Research on assemblages may be characterized by an emphasis on their "expressive potential" (Deleuze & Guattari, 1988) while accounting for the relative stability of (some) assembled orders. A concern with the exteriority of relations means that assemblage thinking is simultaneously concerned with the agency of component parts. This provides a way of describing how actors within the assemblage may possess different resources and capacities to act. Deleuze and Guattari use the notion of the "operator" or "assemblage converter" to highlight the catalytic impact of well-placed elements in either transforming assemblages or ensuring that relations and parts remain stable (1988, pp. 324–325).

Situations create a variety of interests (or stakes) for the component parts (or stakeholders) of an assemblage. Thus, component parts may be defined as stakeholders in a given situation. Although stakeholder theory primarily focuses on the management of companies and their operative environments (Carroll & Buchholtz, 2012; Fassin, 2009; Freeman, 2010), I argue for a broader scope in which a stakeholder framework can be used in concert with assemblage thinking to shed light on stakeholder relationships in particular situations, such as the extractive energy landscapes of Alberta, where different component parts have different interests or stakes in the governance of contested energy spaces.

Stakeholders are defined as directly or indirectly critical to the goals of other actors, be they companies (with the goal of earning profits from extraction) or a local government (with the goal of attracting foreign investments, increasing corporate taxes and creating jobs) or communities (with the goal of obtaining employment, controlling their resources, receiving compensation, gaining welfare services or having a safe environment) (Freeman, 2010). Second, they are directly or indirectly affected by other actors' operations, and they form the social and geographical framework within which the actors operate (Freeman, 2010). Stakeholder management binds the actors to each other and to the geographical context, because the stakeholders differ according to geographical area, and stakeholder agendas are context specific.

"Stakeholder theory is about managing potential conflict stemming from divergent interests" (Frooman, 1999, p. 193). To complete the picture of stakeholders and their influence on other actors, it is fruitful to identify stakeholders' properties and their ability to act. Mitchell et al. (1997) create an in-depth version of stakeholder theory, which they present in a model based on three stakeholder attributes or capacities: stakeholder *strength*, *legitimacy* and *urgency* (see Chap. 7 for details).

The claim that power is produced or generated by groups or institutions distinguishes this concept from that of power as an inscribed capacity. Power is produced by a process of mobilization whereby firms or organizations—for example—reflect upon their own resources to achieve certain goals, and realizing their limitations, attempt to pool their resources with like-minded organizations as a means of securing what is now a common goal (Allen, 1997). This resonates well with DeLanda's notions of assemblages with independent components but with relations that are exterior to their terms.

Power, in this context, is a fluid medium. It is also closely related to resources. The term "resources" in this context implies both human resources—capacities and competences—and material resources, including financial resources. Resources are the medium through which power is exercised. Thus, power is the ability to mobilize resources and to use them to secure specific outcomes (Giddens, 1979, p. 91). Accordingly, when pooled collectively, resources can actively empower groups and organizations, but only as long as such resources are used in concert (Allen, 1997; Arendt, 1986). Component parts may occasionally align, or pool their resources as a means of securing common goals (Allen, 1997, 2003, 2011a, 2011b). Drawing on Foucault, I claim that a contested energy space is an archipelago of different powers (Foucault, 2007, p. 156).

To summarize the conceptualizations of contested energy spaces, I have shown that contested energy spaces are assemblages of entities or components that are situated both inside the extraction site (such as local communities, company branches or NGOs) and outside (such as company HQs, national government or NGOs). The components are scaled, as they differ in the extent of their influence (such as local governments, national governments or global companies). Moreover, the components are most certainly networked, as they organize and assemble in different semi-structured ways to extend their influence. This networking is characterized by an ephemeral nature and emerging structure, in that some of the confluences of interests or arguments are more or less coincidental and sometimes even spurious.

References

Allen, J. (1997). Economics of power and space. In R. Lee & J. Wills (Eds.), *Geography of economies*. London, UK: Arnold.

Allen, J. (2003). *Lost geographies of power*. Oxford, UK: Blackwell Publishing.

Allen, J. (2011a). Powerful assemblages? *Area, 43*(2), 154–157. https://doi.org/10.1111/j.1475-4762.2011.01005.x

Allen, J. (2011b). Topological twists: Power's shifting geographies. *Dialogues in Human Geography, 1*(3), 283–298. https://doi.org/10.1177/2043820611421546

Anderson, B., Kearnes, M., McFarlane, C., & Swanton, D. (2012). On assemblages and geography. *Dialogues in Human Geography, 2*, 171–189. https://doi.org/10.1177/2043820612449261

Arendt, H. (1986). Communicative power. *Power, 64*.

Bridge, G. (2001). Resource triumphalism: Postindustrial narratives of primary commodity production. *Environment and Planning A, 33*(12), 2149–2174.

Bridge, G. (2004). Contested terrain: Mining and the environment. *Annual Review of Environment and Resources, 29*, 205–259.

Callon, M. (2008). In T. Pinch & R. Swedberg (Eds.), *Economic markets and the rise of interactive agencements: From prosthetic agencies to habilitated agencies* (pp. 29–56). Cambridge, UK: MIT Press.

Carroll, A. B., & Buchholtz, A. K. (2012). *Business and society: Ethics, sustainability and stakeholder management*. Mason, OH: South-Western, Cengage Learning.

Castells, M. (1996). *The information age: The rise of the network society* (Vol. 1). Oxford, UK: Blackwell Publishers.

Catton, W. R. (1982). *Overshoot: The ecological basis of revolutionary change*. Urbana, IL: University of Illinois Press.

DeLanda, M. (2006). *A new philosophy of society: Assemblage theory and social complexity*. London, UK: A&C Black.

DeLanda, M. (2016). *Assemblage theory*. Edinburgh, Scotland: Edinburgh University Press.

Delaney, D., & Leitner, H. (1997). The political construction of scale. *Political Geography, 16*(2), 93–97.

Deleuze, G., & Guattari, F. (1988). *A thousand plateaus: Capitalism and schizophrenia*. London, UK: Bloomsbury Publishing.

Deleuze, G., & Parnet, C. (2007). *Dialogues II*. New York, NY: Columbia University Press.

Dicken, P., Kelly, P., Olds, K., & Yeung, H. W.-C. (2001). Chains and networks, territories and scales: Towards a relational framework for analysing the global economy. *Global Networks, 1*(2), 89–112.

Escobar, A. (2007). The 'ontological turn'in social theory. A commentary on 'Human geography without scale', by Sallie Marston, John Paul Jones II and Keith Woodward. *Transactions of the Institute of British Geographers, 32*(1), 106–111.

Fassin, Y. (2009). The stakeholder model refined. *Journal of Business Ethics, 84*(1), 113–135.

Ferguson, J. (2005). Seeing like an oil company: Space, security, and global capital in neoliberal Africa. *American Anthropologist, 107*, 377–382.

Foucault, M. (2007). The meshes of power. In *Space, knowledge and power: Foucault and geography* (pp. 153–162). Aldershot, UK: Ashgate.

Freeman, R. E. (2010). *Strategic management: A stakeholder approach*. New York, NY: Cambridge University Press.

Frooman, J. (1999). Stakeholder influence strategies. *Academy of Management Review, 24*(2), 191–205.

Giddens, A. (1979). *Central problems in social theory: Action, structure, and contradiction in social analysis* (Vol. 241). Berkeley, CA: University of California Press.

Haarstad, H., & Wanvik, T. I. (2016). Carbonscapes and beyond conceptualizing the instability of oil landscapes. *Progress in Human Geography, 41*, 432. https://doi.org/10.1177/0309132516648007

Hetherington, K. (1997). *The badlands of modernity: Heterotopia and social ordering*. London, UK: Psychology Press.

Howitt, R. (1998). Scale as relation: Musical metaphors of geographical scale. *Area, 30*, 49–58.

Jessop, B., Brenner, N., & Jones, M. (2008). Theorizing sociospatial relations. *Environment and Planning D: Society and Space, 26*(3), 389.

Leitner, H., & Miller, B. (2007). Scale and the limitations of ontological debate: A commentary on Marston, Jones and Woodward. *Transactions of the Institute of British Geographers, 32*(1), 116–125.

Logan, O., & McNeish, J. (2012). Rethinking responsibility and governance in resource extraction. In J. McNeish & O. Logan (Eds.), *Flammable societies: Studies on the socio-economics of oil and gas* (pp. 1–44). London, UK: Pluto Press.

Marston, S. A., Jones, J. P., & Woodward, K. (2005). Human geography without scale. *Transactions of the Institute of British Geographers, 30*(4), 416–432.

McClintock, A. (1995). *Imperial leather: Race, gender and sexuality in the colonial contest*. New York, NY: Routledge.

Mitchell, R. K., Agle, B. R., & Wood, D. J. (1997). Toward a theory of stakeholder identification and salience: Defining the principle of who and what really counts. *Academy of Management Review, 22*(4), 853–886.

Paasi, A. (2004). Place and region: Looking through the prism of scale. *Progress in Human Geography, 28*, 536–546.

Palmås, K. (2007). *Deleuze and DeLanda: A new ontology, a new political economy?* Paper presented at the Economic Sociology Seminar Series, London School of Economics & Political Science.

Redford, K. H. (1991). The ecologically noble savage. *Cultural Survival Quarterly, 15*(1), 46–48.

Sheller, M., & Urry, J. (2006). The new mobilities paradigm. *Environment and Planning A, 38*, 207–226. https://doi.org/10.1068/a37268

Sheppard, E. (2002). The spaces and times of globalization: Place, scale, networks, and positionality. *Economic Geography, 78*(3), 307–330.

Stevens, P., Kooroshy, J., Lahn, G., & Lee, B. (2013). *Conflict and coexistence in the extractive industries.* Chatham, NJ: Chatham House. Retrieved from www.usa.com/frs/chatham-house-publishers-inc.html

Tsing, A. L. (2005). *Friction: An ethnography of global connection.* Princeton, NJ: Princeton University Press.

Xiang, B. (2013). Multi-scalar ethnography: An approach for critical engagement with migration and social change. *Ethnography, 14*(3), 282–299.

Chapter 3
Reflections on the Research Process

Abstract Every research process is saturated with methodological reflections, trials and tribulations and ultimately—decisions. In this chapter, the author delves into the challenges and choices that structured the research behind this book: what to look for, where to look for it, how to produce and process qualitative data, and how to distil new knowledge and new theory through analysis. By combining Grounded Theory Method (GTM), personal sensitizing concepts, situational mapping and analysis, and multi-sited fieldwork, this chapter gives a unique and intimate insight into the research process of qualitatively oriented geography.

Keywords Grounded theory method · Situational analysis · Sensitizing concepts · Multi-sited fieldwork · Constructing theory

I consider myself to be a qualitatively oriented researcher. Qualitative research is a situated activity that locates the observer in the world. Qualitative research consists of a set of interpretive, material practices that make the world visible. As a qualitative-oriented researcher, I turn the world into a series of representations, including field notes, interviews, conversations, photographs, recordings and memos to myself, involving an interpretive, almost naturalistic approach to the world (Denzin & Lincoln, 2000). This means that qualitative researchers like me study things in their natural settings, attempting to make sense of, or to interpret, phenomena in terms of the meanings that people bring to them.

However, it is understood that each research practice makes the world visible in a different way. "Choice of research practices depends upon the questions that are asked, and the questions depend on their context" (Nelson, Treichler, & Grossberg, 1992). Hence, the research process is a perpetual, emergent dialogue between the researcher and the world. Looking back at the research behind this book, I realize how messy, creative and incremental the research processes are and how decisive the research process is for a researcher's conceptualizations of the world. From my early memos and field diary notes, I have rediscovered the curious and exploratory nature of a researcher's first encounters with the field of inquiry, and how these encounters have formed both my research and myself along the way.

The thematic focus of this book is a contested field of research, involving a multitude of interpretations and meanings assigned to different actors and agitators. I ask why some Indigenous communities support extractive industry developments on their traditional territories despite substantial destruction of the local environment and traditional Indigenous land use practices? I investigate (a) how to *conceptualize* the *socio-material complexity* of contested energy spaces, (b) how to identify *instability* and *potential for change* in contested energy spaces and (c) how to *understand the power play* between industry, state and indigenous communities in the contested energy spaces of the Canadian North, and I explore the *impact of non-human agency* within the assemblage of contested energy spaces.

I have striven to deploy several interconnected interpretive practices, hoping to improve my understanding of the relevant subject matter (Denzin & Lincoln, 2000). At the same time, it was important to me to choose a methodology and research method with which I as a researcher and the participants of my research would all feel comfortable, given that the data are always produced in concert by the researcher and the researched (Charmaz, 2014; Glaser & Strauss, 2009).

3.1 Epistemological Choices: What to Look for

Any research project or investigation is built upon a fundamental epistemological stance (Holt-Jensen, 2009), guiding the formulation of research problems, the evaluation of theory, the choice of appropriate techniques and—above all—the interpretation of results. How do we know what we know, and how do we record knowledge about the world? These are tempting and grand questions to ask at any level of a research endeavour.

In epistemological discourses, the sources of knowledge and justification of knowledge are for the most part tightly connected to our human cognitive faculties. In general epistemology, these sources are often referred to as: (a) perception, (b) introspection, (c) memory, (d) reason and (e) testimony (Steup, Turri, & Sosa, 2013). However, to trust our knowledge and beliefs, we must consider the reliability of these faculties and our subsequent ability to justify our knowledge claims.

In the research for this book, I consider why certain groups or Indigenous people support industrial developments on their traditional territories. I have approached this issue by mapping the complex power structures of the contested energy space of Wood Buffalo, Canada. When interviews, conversations, participatory observations and texts are used as sources of knowledge, most of the aforementioned faculties are activated, either by me as researcher or directly or indirectly by my informants. However, ultimately, as a qualitatively oriented social scientist, I am disproportionally dependent on testimony—from individuals, texts, representatives and institutions. Testimony differs from the other sources of knowledge because it is not distinguished by having its own cognitive faculty (Lackey & Sosa, 2006).

Testimonies, I would claim, are in turn dependent on the other four faculties: the perceptions, introspection, memories and reasoning of my informants and me. With all the contingencies related to these faculties, there will always be a possibility of errors, mistakes, fraud and lack of honesty among interviewees, informants and other sources. Hence, one might say that the justification of our knowledge claims must be based on an interdependent set of faculties that all have flaws and fallibilities (Steup et al., 2013).

As a social science researcher, I believe that a way of managing these fallibilities is to employ different but interconnected forms of methodological triangulation: (1) methods (faculty) triangulation, (2) source triangulation and (3) stakeholder triangulation. I have striven to employ different ways of encountering and engaging with my informants (interviews, conversations, observations and participatory processes). I have activated a wide variety of sources (strategy papers, traditional land use studies, individual testimonies, institutional statements, juridical verdicts and group discussions), and I have engaged a variety of stakeholders (representatives of indigenous groups, consultants, corporate managers, public officers and bureaucrats). These efforts were conducted to meet the scholarly obligation to seek clarity and to justify knowledge claims while gaining a wider understanding of the issues: the complex power structures of contested energy spaces in Wood Buffalo, Alberta.

3.2   The Situation as a Case Study: Where to Look for it

The study behind this book should be treated as a case study. To define a case, Yin (2013) suggests that the term refers to an event, an entity, an individual or even a unit of analysis. It is an empirical inquiry that investigates a contemporary phenomenon in its authentic context using multiple sources of evidence. Aase and Fossåskaret (2007) see case studies as being concerned with how and why things happen, allowing the investigation of contextual realities. Case studies can be useful in capturing the emergent and immanent capacities of life in situations. They enable the researcher to gain a holistic view of a certain phenomenon, although the data produced in cases are always to some extent extracts from an all-encompassing complexity. According to Yin, there are three types of case study research: exploratory, descriptive and explanatory (Yin, 2013). In this study, I argue that the different varieties somehow overlap, in that I have sought to explore, to describe and to explain practices among various stakeholders in the contested energy spaces of the Canadian north.

This study treats communities, companies and institutions as entities and objects of study in their own right. This implies that I have mainly interacted with the management level, with the exception of suppliers of consultancy services, where I have been in dialogue with each consultant. This is not to disregard the fact that these entities are plural, fragmented units containing divergent voices (assemblages within assemblages); it is only a way to limit my scope to avoid becoming stuck in a nihilist, deconstructivist quagmire. According to Manuel DeLanda, assemblages of social and material component parts (like companies or communities) are unique and singular entities. This notion of the "individual" requires brief clarification; as a realist,

DeLanda argues that the word "individual" can be applied to anything, including communities, organizations, atoms, folded pieces of paper, species and ecosystems. As he puts it, every assemblage is an individual singularity with its own contingent history of emergence and conditions of pertinence (DeLanda, 2006). One of the major contributions of DeLanda's assemblage thinking is that it serves as a framework of micro- and macro-levels of social reality. Interactions between people yield institutional organizations, interacting organizations yield cities, interacting cities organize the space in which nation-states emerge and so on. In assemblage thinking, wholes can serve as components of larger assemblages. The possibility of linking the micro- and macro-levels of social reality in this way is the result of recognizing that social processes occur on more than the two levels of micro and macro. By introducing intermediate levels of scale, assemblage theory can build up from the smallest entity (like an individual person) to increasingly larger assemblages (DeLanda, 2006).

3.3   Grounded Aspirations

Throughout the research for this book, I have been both consciously and subconsciously inspired by grounded theory methods (GTM) for a number of reasons. First, I have sought a way to systematize the collection and production of data. This was particularly important because of the messiness and contested realities of the field. Grounded theory provided me with a more systematic approach to data production, and this has given me a quicker overview and a more thorough insight into the contested energy space of Wood Buffalo, Alberta, Canada. Second, my early observations were inconsistent with most of the existing literature on the subject; neither were they in line with conventional theories of Indigenous responses to industrial developments. Hence, I sought a method of generating empirically based (grounded) theories based on these findings and subsequently enabling me to contribute to the ever-expanding literature on contested energy spaces. Grounded theory inspired me to embark boldly on a theory-constructing journey. Finally, I sought ways to construct empirically based theories on subaltern agency that may be tested elsewhere, in other settings, to analyse subaltern agency beyond resistance and reaction.

GTM is based on the diverse backgrounds of the two originators—Anselm Strauss and Barney Glaser. Stated simply, grounded theory methods consist of systematic yet flexible guidelines for collecting and analysing qualitative data to construct theories "grounded" in the data themselves (Charmaz, 2014). The research process brings surprises, sparks ideas and hones the researcher's analytic skills. The guidelines offer a set of general principles and heuristic devices rather than formulaic rules (Atkinson, Delamont, & Coffey, 2004). First, grounded theory is an empirical approach to the study of social life through qualitative research and analysis. In this method, the analyst initially codes the data, giving temporary labels (codes) to particular phenomena (Clarke, 2005). Unique to this approach has been the fact that analysis begins as soon as there are data. Coding begins immediately, as does theorizing based on that coding, however provisionally (Glaser, 1978).

Methods alone—whatever they may be—do not generate good research or astute analyses (Charmaz, 2014). Mechanistic applications of methods yield mundane data and routine reports. A keen eye, open mind, discerning ear and steady hand can bring the researcher close to the object of study and are more important than developing methodological tools (Charmaz & Mitchell, 1996). This creative, pragmatic approach to methodology and research appeals to me and has been a guiding principle throughout the research behind this book.

3.4 Pragmatic Heritage and Hybridization

Pragmatism pervades the grounded theory method, and Charmaz (2014) claims that this is because of Strauss's Chicago school heritage. Strauss viewed human beings as active agents in their own lives and in their worlds rather than as passive recipients of larger social forces. Strauss brought notions of human agency, emergent processes, social and subjective meanings, problem-solving practices and the open-ended study of action to grounded theory (Charmaz, 2014). He particularly underscored the American pragmatist James Dewey's (1916) view that disavowed what he termed "the spectator theory of knowledge". Rather, we are part of the world that we study and the data that we collect (Bryant, 2009). Dewey proposed "the experimental theory of knowledge", where all knowledge is seen as provisional and is judged in terms of its usefulness for the knowing subjects, placing action and emancipation at the centre (Bryant, 2009). Hence, in all research practice, personal preconceptions should be considered. In line with this pragmatic approach, I have been inspired to propose a hybrid variety of grounded theory by invoking theory development strongly inspired by earlier theories around assemblages, power and pragmatism through active engagement with my personal preconceptions and former experiences as a practitioner.

3.5 Personal Preconceptions and Sensitizing Concepts

Gadamer describes how our background shines through what he calls our preconceptions (Gadamer, 2006; Rennie, 2000). Charmaz (2014) underscores that every researcher holds preconditions that influence, but may not determine, what we attend to and how we make sense of it. My own preconceptions were formed by a myriad of past experiences, but I will limit myself to describing some of my formative professional work during the past 20 years.

For most of my professional life, I have found myself to be an intermediary, a negotiator and translator of interests. First, as a young activist,[1] the feeling of fighting superior powers was both an inspiration and a reason for being. Second, serving

---

[1] Early political awakening through student-driven solidarity work/humanitarian work.

as a junior manager heading a progressive, young and ambitious political advocacy division in an established, large humanitarian NGO,[2] there were constant negotiations with senior management and the board. Third, my decades-long engagement with non-profit, non-governmental organizations[3] in development aid and political advocacy has left me with a strong social consciousness and a notion of social justice that has served as a guiding principle in this research process. Finally, when I was communication consultant and senior manager in a design company,[4] negotiations with demanding clients, ambitious employees and impatient owners were part of everyday life.

GTM has provided me with a set of analytical approaches that are tools both to discover and to utilize my preconceptions of social justice, activism and the pragmatic nature of negotiations. GTM can move social justice studies beyond description while keeping them anchored in their respective empirical worlds. Thus, grounded theorists can offer integrated theoretical statements about the conditions under which defined forms of injustice or justice develop, change or continue (Charmaz, 2014). I have tried to recognize my preconceptions and to turn them into assets for my research. One way of accomplishing this has been to treat them as sensitizing concepts. An active pragmatic focus on social justice has sensitized me to view both large collectives and individual experiences in new ways.

Blumer's (1980) notion of sensitizing concepts is useful at this juncture. A sensitizing concept is a broad term without definitive characteristics; it sparks the researcher's thought about a topic. Charmaz encourages researchers to treat these concepts as points of departure for studying the empirical world (Charmaz, 2014). Sensitizing concepts can help the researcher to start to code and processing data. These concepts provide starting points for initiating the analysis but do not determine its content. My sensitizing concepts were constructed around key questions derived from my previous experiences as negotiator/mediator: How do social processes and situations change? How is power mobilized, with preferential treatment of hidden, unknown or unlikely power mobilization among subaltern groups? What meanings do participants attribute to the process or situation? What do they emphasize, and what do they leave out? In Sect. 3.8, I elaborate on how my personal preconceptions have influenced and coloured my analysis and conceptual framing of the research topic.

3.6   Production of Data: Multi-Sited Field Research

Rather than being excavated, found or collected, data are produced—in the encounter between the researcher, the study participants, and the material and discursive environments constituting the field. Glaser (2002) states that "all is data". Yes,

---

[2] Norwegian Church Aid, acting as head of campaign and media division.

[3] Norwegian Peoples Aid, Norwegian Red Cross, Norwegian Refugee Council, Norwegian Church Aid, Save the Children Norway, Operation Day's Work and Children at Risk Foundation.

[4] Orangeriet As and NORPR, both part of Media Bergen Group.

everything I have learned in the research setting(s) or about my research topic can serve as data. However, data vary in quality, relevance and usefulness for interpretation (Charmaz, 2014). Social sciences today face a common challenge—an increasingly tangled and complex world (not necessarily diverse—rather homogeneous, but complex in its structure), where the social sciences are in the process of standardizing the tools for deciphering and analysing this reality. The notion that global diversity would be identified and understood through minute and slow studies of small communities is gradually being replaced by a complex world of interconnected and overlapping networks, relationships, meeting points and breaks. Such a world cannot be explained or analysed from a single place or community (Næss, 2015; Wanvik, 2015). We should attempt to distinguish and systematize connections, patterns, correlations and disruptions, acts and reprisals (Therborn, 2011, p. 208). Therefore, I have found myself travelling between corporate headquarters in Oslo and Calgary, organizational hubs in Edmonton, Lac La Biche and Fort McMurray, and local community headquarters and individual trap lines in the midst of the boreal forests of Wood Buffalo, analysing how the parts fit together.

In line with the theoretical framing of this book (see Chap. 2), the contested energy spaces stretch far beyond the borders of the geographical place of extraction. Several disciplines note the importance of multiscalar approaches to practical research in the social sciences. According to Xiang (2013), multiscalar research is primarily concerned with how social phenomena are constituted through actions on different scales. Howitt and Stevens (2005) underscore multilocational fieldwork in locations linked by various aspects of contested energy spaces (e.g. networks: common ownership, downstream process integration, competition or government policies). This implies that the *local* quickly emerges as a set of particular kinds of relationships and networks linked to a much wider set of scale relationships rather than a singularity focused on a bounded location. The research process brought me to head offices in Calgary and Edmonton, to corporate headquarters in Oslo and all the way down to the trap lines of Métis elders. Any *local* is always and inescapably contextualized by a range of critically important power relations that were constructed on several scales. Relations across multiple scales and networks provide a vantage point from which to understand how multi-sited connections actually work and what the sites mean to each other (Xiang, 2013).

Analyses using assemblage approaches depend on the careful tracing of actor networks where many players (human and non-human) are at work constituting the outcome. However, reconstructing histories of ideas, stakeholder relations and political configurations through these interactions of objects/actors does not mean that explanations freely assemble a limitless cast of characters, each equal in power, all equally efficacious and all with coherent intentions. Rather, it entails making "this issue of power and agency a question, instead of an answer known in advance. It means, acknowledging something of the unresolvable tension, the inseparable mixture and the impossible multiplicity out of which intension and expertise emerge" (Mitchell, 2002, p. 53).

However, the complexity involved in writing assemblage geographies can invite sloppiness. The Deleuzian injunction to "begin in the middle" methodologically, which admirably embraces the reality that all actors and subjects are always already in the middle, is important for assemblage geography but is also ripe for abuse. Half-finished stories, endings that become beginnings and flashbacks are all part of the arsenal of excellent authors, but they can also be the signatures of unfinished research, quick editing and incomplete thinking (Robbins & Marks, 2010). Therefore, finding the relevant samples in this case has been a continuously challenging task, pushing me to seek precision, thoroughness and accountability, supported by the GTM framework.

Sampling in GTM is driven not necessarily (or not only) by attempts to "represent" some social body or population or its heterogeneities but also especially and explicitly by theoretical concerns. Theoretical sampling has been integral to grounded theory from the outset and remains a fundamental strength of this analytic approach (Clarke, 2005). In GTM, there is an explicit recognition of the iterative relationship between data gathering/production and analysis, and the ways in which they interrelate and guide each other. By extension, this more mindful and insightful view of the iterative process of data gathering-cum-analysis explains the idea of *theoretical sampling* that is key to GTM (Bryant, 2009). For some critics of GTM, it might seem to be a case of researchers looking for confirmation of their initial ideas, as opposed to trying to falsify or disprove them, but from the pragmatist point of view, it is far more a case of seeking ways in which my emerging concepts actually *work* in elucidating the specific research context (Bryant, 2009).

In this study, sampling has been driven not only by attempts to represent some social body or population but also especially and explicitly by theoretical concerns that have emerged in the provisional analysis. Such theoretical sampling focuses on finding new data sources (persons or things) that can "best explicitly address specific theoretically interesting facets of the emergent analysis" (Clarke, 2003, p. 557).

In line with Silverman's recommendation, the theoretical apparatus that I use has informed my sampling and my choice of case. Sampling in qualitative research is neither statistical nor purely personal; it is, or should be, theoretically grounded (Silverman, 2013). Theoretical sampling and purposive sampling are often treated as synonyms. Indeed, the only difference between the two procedures applies when the purpose behind "purposive" sampling is not theoretically defined (Silverman, 2013). Bryman argues that qualitative research follows a theoretical, rather than statistical, logic. The issue should be couched in terms of "the generalizability of cases to theoretical positions rather than to populations or universes" (Bryman, 1988, p. 90). Theoretical sampling means selecting groups or categories to study on the basis of their relevance to the research questions that I wish to explore, to my theoretical position, and most importantly to the explanation or account that I have developed (Mason, 2002).

In this research, I have combined different sampling methods. In the early stages, I relied on purposive sampling to gain pace. Through corporate contacts and literature reviews, I identified some vital gatekeepers. These sampling efforts resulted in some version of snowball sampling, as my initial contacts were able to point me in the direction of other valuable sources. As I approached the situation through contacts in industry, private consultancies, public institutions and local communities, I could "triangulate" my sampling in a way that I hoped would compensate for otherwise problematic issues of bias in relation to snowball sampling, favouring one version of the situation.

It was important for me that the interviewees could speak on behalf of the company, institution or community that they represented. This is important for data collection and analysis, because the social and institutional position of each informant is essential when analysing the implications of his/her statements (Haarstad, 2009). Giampietro Gobo shows that the qualitative researcher should focus the investigation on interactive units such as social relations, encounters and organizations (Gobo, 2008). The sampling had to be developed during the research period, rather than beforehand, because this was a new situation for me, and I was reliant on "insiders" to gain access. The sample is defined not so much by the size of the universe of possible respondents as by the *situation* or event, and the movers and shakers in that situation.

I believe that through theoretically informed purposive sampling based on stakeholder identification and situational analysis, I was well equipped to describe and understand the tensions and transactions between state actors, companies and communities in the contested energy space of Wood Buffalo. This was not so much to describe a whole class of phenomena or actors (Becker, 2008) as to understand their situation.

3.7 Fieldwork: A Variety of Methods, Sources and Stakeholders

I entered the field to understand the experience of Norwegian multinational companies with the situated politics of place in the energy spaces of Canada. However, it soon became apparent that the complexity of the local politics demanded a more nuanced focus than one on the corporate sector alone. Fieldwork can be considered to mediate between theory, method and data; new information makes the researcher adopt new theories, concepts or methods, which in turn produce new information (Wadel, 1991). For a researcher, access to contested and conflictual situations can be challenging. In my case, my initial contact was through industry and the regional scholarship for research on traditional land use practices and colonial history.

Key Informants

Researching contested energy spaces demands a great deal of political sensitivity, and it has proven to be challenging to produce data in Canada, where the polarization between supporters and opponents of oil sands extraction is pronounced and where conflict cuts across several political, economic and cultural aspects. In addition to a large amount of secondary literature on topics such as consultations, impact assessments and compensation agreements, I have been dependent on key informants who either have given me in-depth knowledge of the various aspects of this complex problem or have acted as gatekeepers and opened doors to other, less accessible sources and stakeholders.

From the fieldwork of previous research,[5] I had established some contacts within the Norwegian state-owned oil company Statoil and its management in Indonesia. Through these contacts, I reached Statoil Canada and their CSR director. My main contact was a Norwegian expatriate who was willing to share the corporate social responsibility strategies and some insider reflections on local content and compensation schemes. He was attentive but somewhat aloof and referred to the company's Canadian lawyers and communications consultants, who did not appreciate researchers or activists, especially "Norwegians who came to sniff" (Statoil Canada manager, Skype interview, 2014). Unfortunately, he resigned shortly thereafter: first from the Canadian company and later from his position in Bergen. However, I met him on several occasions, before, during and after he ended his involvement in Statoil, and he has been very helpful with various sources and background documentation. My contact also recommended me both to the Statoil Headquarters in Oslo and to the newly appointed Canadian Director of CSR in Statoil Canada. However, I was soon to discover that the welcoming tone of Norwegian expatriates had turned into a more apprehensive attitude among Canadian nationals.

Given that the contested energy spaces of Canada were a new experience for me, I decided to embark on a pre-field visit in March 2014 to gain an overview of the field, to explore the geography of the situation and to start building a network that might be useful when I returned the following year. I had made some arrangements with Statoil, some multi-stakeholder organizations and some regional scholars, lawyers and consultants, and I felt well-prepared. However, during the actual pre-field trip, I found myself in a situation resembling a 90-min interrogation at the corporate headquarters of Statoil Canada, where my ambitions and intentions were questioned and scrutinized, and where several of the planned meetings were suddenly changed or cancelled. The level of caution and reluctance from the company came as a surprise to me, and I was soon to discover that the conflictual nature of previous encounters with Norwegian activists who had left the company, and especially the Canadian national employees and management, nervous and not particularly eager to speak to a Norwegian researcher (Fig. 3.1).

Later, I found that the company had followed my tracks, gathering information about my whereabouts and my efforts to establish contact with local community representatives in Conklin and with multi-stakeholder initiatives. There are strong indications that the company partially discouraged stakeholders from participating in my study. At one point, I was told bluntly: "Statoil is an important member of (a multi-stakeholder organization) and would prefer you to use other resources to do your research" (email correspondence 2014). This forced me to seek other options

---

[5] Research project on CSR and Norwegian enterprises in Indonesia 2011–2013.

Fig. 3.1 Early field trip 2014. Statoil headquarters in Oslo. Photo by the author

to contact local stakeholders. Here, the literature review and dialogue with regional scholars proved to be fruitful.

In an early phase, while seeking scholars and professionals working for or with local communities, I came across a consultant and historian with extensive connections among the local communities around Fort McMurray. I was very pleased to learn that having worked with all three of the communities in this study, and serving on several boards of multi-stakeholder organizations in the area dealing with indigenous peoples, environment and the oil sands, he was willing to discuss these issues with me as a researcher and colleague.

I gained access to all the communities, with the help of my key informant. Sometimes he opened doors, sometimes he vouched for me and sometimes he personally was a valuable source of information. Although this entry into the field may be criticized for being too dependent on individuals, it is difficult to imagine how I could have succeeded otherwise, given the challenging and conflictual nature of the situation at hand. I was fortunate to spend considerable time with my key informant and his extended network of community elders, leaders, consultants and managers, and these contacts were instrumental in framing and limiting the scope of my research.

A Variety of Sources

My pool of sources was determined by access. Given that contested energy spaces in general are difficult to access, some sources are more difficult to acquire than others. The data sources for this study were the following.

- Environmental Impact Assessments (EIA) from two companies totalling 5000–6000 pages of information about the environmental and social implications of oil production in the field from Station Canada and Tech Industries.
- Four traditional land use studies from Fort McKay Métis, McMurray Métis, Conklin Métis and Chipewyan Prairie First Nation.
- Extensive secondary literature on consultation processes, environmental impact assessments and impact benefits agreements.
- Strategic documents, annual reports, CSR reports and action plans from Statoil.
- Area plans, consultation strategies, guidelines for impact assessments and the six most significant Supreme Court verdicts handed down in relation to indigenous rights and opportunities related to contested energy spaces.
- News stories, blogs, YouTube interviews and newscasts related to indigenous management of multinational oil companies, traditional way of life, conflicting land use and conflicts.
- Email correspondence and Skype calls with interviewees and comments by ten key players in the field.
- Semi-structured interviews and conversations with 30 key players from indigenous organizations, consultants, representatives for indigenous companies, multi-stakeholder organizations, local chambers of commerce and local authorities.
- Discussions with scientists from the field.
- Observations from stakeholder gatherings, public seminars and cultural events.

Field visits were conducted in three stages, with preparatory fieldwork in June 2014, a month of fieldwork in June 2015 and a post-field trip in September 2016.

Data Production Methods

I have been fortunate to engage with company managers, community elders and leaders, and municipal employees in a wide range of formal and informal settings. The observations were both planned and coincidental, but I have always kept thorough notes of all activities, so I label them *formalized observations*. In line with contemporary trends in qualitative research, I have conducted in-depth interviews but of a somewhat special kind in that I have coined the term "camp-fire conversations", akin to participatory observations or dialogues. These conversations occurred irregularly throughout my three field exposures and constitute the backbone of my data material.

Typically, these conversations were initiated by the elders on an irregular basis but so frequently that the events may be called regular gatherings of elders and other members of the Métis communities (Fig. 3.2).

In addition, I conducted both traditional semi-structured interviews and less formal face-to-face conversations with consultants, corporate CSR officers, municipal

Fig. 3.2 Camp-fire conversations with Métis elders and youth 2016. Photo by the author

bureaucrats and community leaders and managers. Some of my interviews had to be conducted by Skype or email because of distance.

Organizational documents were an important part of my data material and are evidence in their own right of the ways in which organizations seek to present themselves to particular audiences. That is, the goal of including such evidence in research design is not to triangulate the truth of interview or ethnographic data but rather to provide an additional data source about the case. In line with Charmaz's recommendation, I treated each text as coherent simply because it was there, lying before me as a unified object (Charmaz, 2014). Environmental impact assessments were treated as sources in their own right, as were traditional land use studies, internal corporate strategy documents, annual reports, area plans and consultation strategies, and guidelines for impact assessments. Supreme Court decisions were treated as background information.

Distinctions are made between *personal* and *institutional,* and between *confidential* and *public* sources. During my research, I encountered all varieties. Two types of documents were particularly important to this study. (1) Environmental impact assessments (EIAs) are produced to meet regulatory requirements set by the province of Alberta. This objective has important implications for the format and content of these massive documents. (2) Traditional land use studies (TLUs) are particularly interesting because of the way in which they are produced and their objective. I was fortunate to participate when consultants conducted interviews for such TLUs among Métis elders. TLUs become institutionalized narratives of traditional land use practices, and their purpose is to gain recognition and political influence over particular territories (Fig. 3.3).

Fig. 3.3 TLU assessment, McMurray Métis trap line 2016. Photo by the author

If grounding in the data is crucial, then there is no reason why extant literature cannot be part of that data. GTM may be criticized as an excuse for avoiding a proper literature review. I have deliberately worked to avoid this criticism by treating previous literature in the field of contested energy spaces as data in its own right. One can never enter a research area with an empty head; one can attempt to maintain an open mind, but sometimes it is precisely one's prejudices—in the sense of prior judgements—that provide a basis upon which innovative insights can be developed (Bryant, 2009). It was crucial for me to understand that I inevitably take an active role in the process of "data gathering"; this is not simply a matter of harvesting something that occurs naturally. For pragmatists, reality is always in the making, and data are "carved out" from reality by social actors; this activity is socially located and is not simply an individual, isolated accomplishment (Bryant, 2009).

3.8 Transforming Data into Codes and Categories

Coding relies on solid data, but when are data solid? I had limited opportunities to record conversations, interviews and observations. This was particularly true during camp-fire conversations, both because of time limitations on recording (the camp-fire conversations normally lasted for up to 4 h) and because the recording conditions related to sound quality across a camp-fire were suboptimal. In addition, recordings would have altered the flow of the conversation, disrupting the free flow and sharing of ideas and aspirations. However, all conversations, interviews and observations were transcribed and described immediately after the event, and I coded the data based on my immediate recollections of conversations in field notes and handwritten notes. Coding in the field was performed by writing marginal key words and comments—associations and reflections emerged as I read through my field notes every evening. These key words and comments were later transformed into codes and categories when I entered my data into software[6] designed for qualitative data processing. The codes and categories served as nodes in the software, and I organized my findings into analytical categories.

As mentioned in the previous section, these codes and categories were derived from two sources: first, they "emerged" from the data and the way in which the data sources appeared to me (Glaser & Strauss, 2009). Correspondingly, the codes and categories hinged on me as a researcher, and my own preconceptions and ideas on the world (see Sect. 3.5). Thus, the codes and categories could be generated from what Charmaz (2014) calls *sensitizing concepts*. Based on my background experiences and my initial research questions, I developed some sensitizing concepts that formed much of my research: these are concepts of *change, uncertainty, instability, pragmatism ("what works") and catalysing agency*. These concepts were derived both from practical experience and from theoretical reflections on, and readings of, assemblage theory (see Chaps. 2 and 5), constituting an ontological approach akin to my perception of reality as perpetual process: always in the making, always unstable, changeable and uncertain. During my data production, these concepts acted as reminders. They reminded me to look for alternative voices, alternative mobilization of power and alternative versions of established, skewed power configurations. In the next section, I elaborate on how this informed my analysis.

3.9   Analysing Data

The data analyses are intended to serve multiple purposes in this research. First, I needed to gain a closer understanding of the contested energy space as assemblage—who were the movers and shakers of the assemblage, and what kinds of stakeholders could I identify? Second, I sought to understand the governing mechanisms of the contested energy space and considered the extent to which I could identify potentials for change. Finally, I was eager to understand how indigenous populations and communities viewed their own situation in the midst of the contested energy space of Wood Buffalo and to explain why not all indigenous groups opposed industrial disruption of their traditional territories.

Situational Mapping and Analysis

Situational mapping and analysis supplement traditional or basic grounded theory with alternative approaches to both data gathering and processing/analysis/interpretation (Clarke, 2005). The situation per se becomes the ultimate unit of analysis, and

---

[6] NVIVO 11.

understanding its elements and their relations is the primary goal (Clarke, 2005). Building upon and extending Strauss' work, situational analysis offers three main cartographic approaches:

- Situational maps that lay out the major human, non-human, discursive and other elements in the research situation of inquiry and provoke analysis of relations among them.
- Social world-arena maps that lay out the collective actors, key non-human elements, and the arenas of commitment and discourse within which they engage in ongoing negotiations—meso-level interpretations of the situation.
- Positional maps that lay out the major positions taken and not taken in the data vis-à-vis particular axes of difference.

Primary, situated maps generated from NVIVO word clouds[7] are chaotic, complex and unstructured collections but nevertheless remain fruitful because they are organized according to word frequencies. Early coding was somewhat dependent on these word clouds to gain an initial understanding of the field. Combined with my first encounters with the field, these situational maps contributed to knowledge formation around the nature of the contested energy spaces.

I collected the data in three distinct phases. The first phase contained most of the written materials, such as environmental impact assessments, strategy documents, annual reports, email correspondence and initial interviews with key stakeholders. The second phase comprised conversations and in-depth interviews. The final round primarily contained conversational data from camp-fire conversations and interviews (Fig. 3.4).

To me, this meshwork of words and conceptualizations representing the contested energy space of Wood Buffalo provided me with some initial ideas about the interrelations between the component parts, or stakeholders. It reveals a close relationship between *industry* (signified by "North American" (later: "Statoil"), "project", "Leismer", "development" and "volume"), *environment* (signified by "water", "river", "environmental", "resource", "sands", "moose", "assessment" and "Athabasca") and *local communities* (signified by "community", "traditional", "consultation", "report", "local", "Métis", "first nations", "people", "McMurray", "McKay", "Chipewyan" and "rights").

Based on these initial findings, I coded my data into comprehensive categories, turning the situational map into a less fragmented assemblage of relevant concepts, as shown below (Fig. 3.5).

Here, I coded the data by organizing them into categories in close dialogue with the transcribed findings. I have weighted the concepts and words according to their frequency and their impact in the sources of information that I encountered in the early phases of my fieldwork. The situational map illustrates how the different aspects of the contested energy space of Wood Buffalo appear to the stakeholders overall. At first glance, the weighting seems to be biased in favour of industrial, extractive com-

---

[7]Based on quantified word count and weighting of the entire body of empirical data collected for this study.

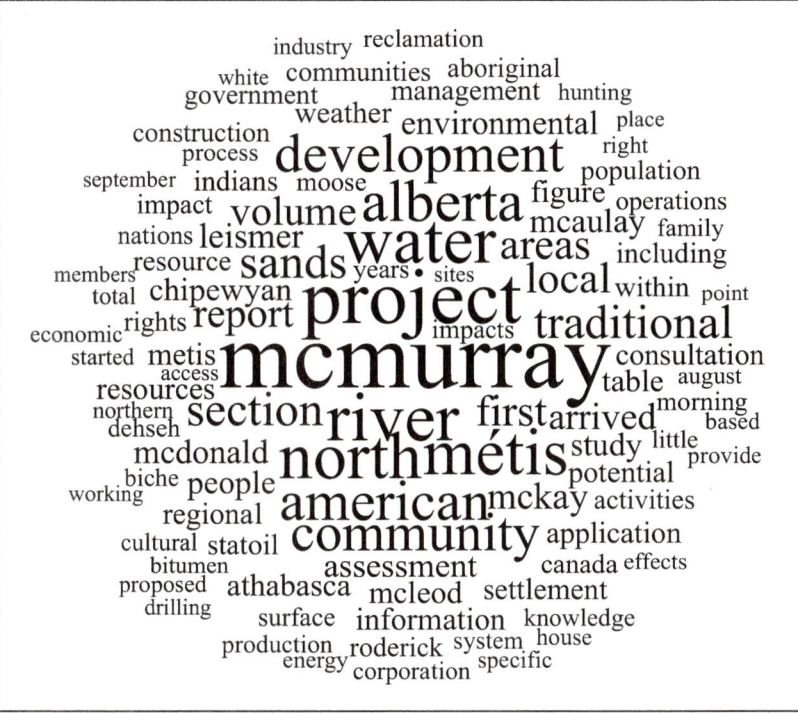

Fig. 3.4 Data processing. Situational map, initial transcripts from phase one data collection. Automatically generated through NVIVO 11

Fig. 3.5 Data processing. Situational map of initial findings, phase one, coded version, generated by the author

Fig. 3.6 Data processing: Situational map 2b: the oil assemblage

ponent parts. Words and concepts such as oil, investments, mining, CSR, money and employment dominate clearly at the expense of more environmental and social concepts such as damage, solidarity, traditions, "Indians" (sic), gathering, fishing, trapping and generations. Early in the process, I identified the primary polarization of the contested energy space of Wood Buffalo as being between proponents (industry and state) and opponents (environmentalists and local communities).

Furthermore, acknowledging the tensions between stakeholders in Wood Buffalo, I experimented with grouping the data into categories, to see whether my understanding of the area changed. This was revealing, pointing towards the polarized realities of contested energy spaces. Here, analysing the wording and conceptualization of the contested energy space from two viewpoints (industry and state versus indigenous communities and environmentalists), two distinct realities materialized in the situation maps below (Figs. 3.6 and 3.7).

These coded situational maps were generated based on all the initial data gathered on my first field trip, together with email correspondence, environmental impact assessments, traditional land use studies and strategy documents from industry. To a large extent, they resonate with the impressions that I gained from meeting state actors, and industry and community representatives.

Industry and other extractive industry proponents such as state institutions focused on job creation, oil opportunities, development, future prospects and growth, regulations and sustainability, and used technical terms such as location, seismic, SAGD, reclamation, water quality and recycling. On the other hand, local communities and environmental NGOs focused more on traditional practices and environmental and social factors such as land, title, prairie, ancestors, trap lines, community, traditions (from earlier times) and nature, while technical aspects were almost totally absent. These findings led me to the preliminary conclusion that there existed a fundamental discrepancy between two distinct ways of viewing the con-

Fig. 3.7 Data processing. Situational map 2b: the environmental assemblage

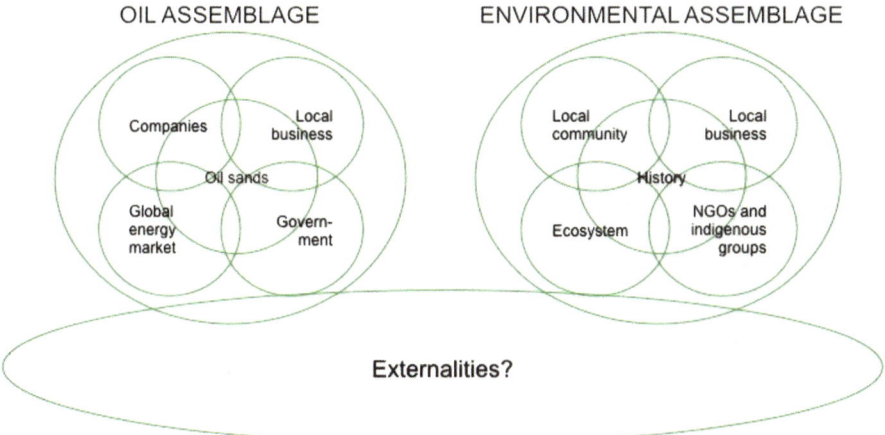

Fig. 3.8 Data processing. Social world map/positional map of energy space of Wood Buffalo

tested energy space of Wood Buffalo. One was represented by the proponents of oil extraction—the oil assemblage—and the other by the opponents—the environmental assemblage (Fig. 3.8).

Turning these situational maps into social world maps was challenging, and I found that these efforts primarily bore fruit in relation to the pedagogical goals of disseminating my research to a wider public. However, they also helped me to identify relevant categories of stakeholders for my second field trip. An early version of this social world/arena map is Fig. 3.8 which shows the differences and similarities of the two assemblages constituting the contested energy space of Wood Buffalo, revealing the overlapping nature of the empirically derived assemblages.

The social world/arena map was combined with a positional map. Here, we find the oil assemblage, containing multinational companies, local businesses, the global energy market and government institutions, tied together by the oil sands deposits and their lucrative profits. This was opposed by the environmental assemblage, comprising local communities, local businesses, local ecosystem services, NGOs and indigenous groups, tied together by historical experiences and traditional land use practices.

However, as I coded and categorized my initial data, there were issues in the material that were not easily reflected in the situational maps, and I became curious as to whether there was more to the contested energy spaces than was apparent to the eye. So far, I had mostly identified power structures and alliances that confirmed the bulk of the literature on extractive industries and local responses. Fuelled by my sensitizing concepts of change, instability and alternative power relations, I embarked on my second field trip, eager to seek alternative understandings of the discourse of contested energy spaces.

Beyond Discourse Analysis

Discourse analysis offers a means of exposing or deconstructing the social practices that constitute a "social structure" and what we might call the conventional meaning structures of social life (Jaworski & Coupland, 2005). My aspiration has been to move beyond my own initial findings in line with conventional discourse analysis, given that the discursive practices of the contested energy spaces have already been scrutinized and recognized as skewed (see for example Clarke, 2009; Huseman & Short, 2012; Le Billon & Carter, 2012; Nikiforuk, 2010). In this study, I sought loopholes in these discourses to identify scope for subaltern groups.

Reichertz (2007) argues that the term "abduction" combines the rational and the imaginative aspects of research, and this is precisely what theoretical sensitivity grounded theory method is meant to encompass: "Something unintelligible is discovered in the data, and on the basis of the mental design of a *new* rule; the rule is discovered or invented, and at the same time, it also becomes clear what the case is. The logical form of this operation is that of abduction. Here one has decided (with whatever degree of awareness and for whatever reasons) no longer to adhere to the conventional view of things" (Reichertz, 2007, p. 219). Charmaz offers another definition of abduction: "a type of reasoning that begins by examining data, and after scrutiny of these data, entertains all possible explanations for the observed data, and then forms hypotheses to confirm or disconfirm until the researcher arrives at the most plausible interpretation of the observed data" (Charmaz, 2014, p. 188).

Critical Foucaultian discourse analysis usually (re)presents and analyses only one discourse in a situation—that with the most power in that situation. Yet, a number of critiques of Foucault have centred on the absence of agency, which allowed the absence of resistance to discourse in his work. By not recapitulating the power relations of domination, I analysed a broader array of discourses to "turn up the volume on the lesser but still present discourses, lesser but still present participants, the quiet, the silent, and the silenced" (Clarke, 2005:xx). This was my objective on the second field trip, when I sought the subtler signs and indications of alternative forms of power formation and mobilization. Of course, with this analytical widening of the range of discourses in a given situation, it becomes even more important to analyse power in that situation carefully (Crampton & Elden, 2007).

One of my early observations was that although Indigenous groups were the obviously weaker component part in Wood Buffalo, these local communities exercised a disproportionate dominance of the discourse. I encountered the relative importance of Indigenous communities and practices in strategy papers, impact assessments and interviews with industry and municipal agencies to such an extent that I developed a keen curiosity about why these miniscule, impoverished communities held such a tight grip on the overall discourse. This impression was strengthened as I conducted interviews and participated in seminars, camp-fire conversations and meetings during my second field trip. To understand this, I began to analyse the governance structure of the contested energy spaces (see Chap. 6). By deciphering the multiscalar discourse of governance applied to the contested energy space of Wood Buffalo, I identified the governance structure of the Canadian oil sands, which is centred around managing the socio-environmental impact of industrial extraction on Indigenous communities. By framing this as a macro-level analysis involving discourses on governance, the post-political condition and corporate social responsibility, I concluded that there was a somewhat skewed power relation between industry proponents and Indigenous communities, but I laid the foundations for a closer examination of the micro-level implications of a predominantly corporate-controlled governance structure (see Chap. 7).

Charmaz (2014) underscores that some of our best ideas may occur to us late in the process and may lure us back to the field to gain an arresting view. As this happened to me, I tried to remain focused on my initial ideas first and to finish one paper or project about them, but I later returned to my data and unfinished analysis in another area. To investigate my ideas of a more diverse and creative mobilization of power among stakeholders in this contested energy space, I decided to return a third time to the field. However, my plans were hampered by a severe wildfire that broke out in the area around Fort McMurray in May 2016, which lasted until June 21 (Fig. 3.9).

The wildfire was devastating, forcing the entire city of Fort McMurray to evacuate its 80,000 inhabitants for days. I had to postpone my trip, but the aftermath of the fire provided a rare opportunity to study the mobilization of power among indigenous communities when I returned in September.

Through a case study of the massive 2016 wildfire (see Chapter 8) in the regional municipality of Wood Buffalo in northern Alberta, I was able to reveal the fragile characteristics of carbonscapes, analysing how assemblage instabilities interfere with conventional ideas of power relations within carbonscapes and how the impact of the fire generated social and political mobilization among local Métis communities of Wood Buffalo and ultimately led to a renegotiation of the assembled order of the oil sands of Alberta.

Fig. 3.9 Métis traditional cabin lost in the fire of 2016. McMurray Métis community. Photo by the author

3.10   Constructing Theory

Theorizing is a practice (Charmaz, 2014). The fundamental contribution of grounded theory methods resides in guiding interpretive theoretical practice, not in providing a blueprint for theoretical products. A constructivist approach prioritizes the studied phenomenon and considers both data and analysis to be created from shared experiences and relationships with participants and other sources of data (Bryant, 2009; Charmaz, 2014; Charmaz & Mitchell, 1996). The analysis becomes more explicably theoretical when a researcher asks what theoretical categories these data represent (pragmatism, instability or uncertainty). Moreover, to the extent that I have interrogated relationships between my categories and fundamental aspects of human existence, such as the nature of social bonds or relationships between choice and constraint, individuals and institutions, or actions and structures, my work becomes yet more theoretical (Charmaz, 2014).

As I have assembled my earlier encounters with the geographical literature, my preconceptions based on former engagements and my data-producing fieldwork, I have slowly developed a theoretical framework that I consider to be promising and fruitful in relation to contested energy spaces. It combines the clear inspiration of DeLanda's assemblage theory (DeLanda, 2006, 2016) with John Allen's power theory on mobilization of resources (Allen, 2003, 2011a, 2011b) and stakeholder theory (Fassin, 2009; Freeman, 2010; Freeman & Velamuri, 2006; Frooman, 1999; Mitchell et al., 1997), supplemented by a pragmatist approach of "what works" (Bryant, 2009; Dewey, 1916; Jones, 2008; Rorty, 1999).

Theories are rhetorical. Theorists attempt to convince readers that certain conclusions flow from a set of premises. The premises on which my assemblage of theories are based can be traced back to my sensitizing concepts (see Sects. 3.5 and 3.8):

instability, change and alternative mobilization of power. A theory can alter one's viewpoint and change one's consciousness. Through my approach to contested energy spaces, I strive to see the world from a different vantage point and to create new meanings for it. I believe that this assemblage of theories is also promising for other geographical subjects of study. However, substantive theories are closely linked to the context in which the research is grounded; only later can these theories become formal, after they have been taken up and used in other contexts and possibly by other researchers (Bryant, 2009).

3.11   Ethics

Working in contested energy spaces is challenging in a number of ways. Obviously, the conflictual character of a situation prompts careful consideration regarding ways to engage with different stakeholders. Ethically informed and considered research first and foremost involves three basic tenets: information, consent and independence (Silverman, 2013). Research subjects must be properly and fully informed about the purpose, methods and intended uses of the research, what their participation in the research entails and what risks may be involved (Silverman, 2013). During all my initial communications and engagement with stakeholder informants, I presented myself as a PhD student from the University of Bergen, and I described the project and my research objectives. A statement of intent and objectives prefaced all formal interviews. Less formal meetings, conversations and participatory activities all included a formal introduction to me as a researcher.

Given the discrepancies between the positions of stakeholders, and the danger of violating confidentiality related to agreements between corporations and communities, I avoided asking for sensitive information regarding specific agreements between my informants and corporations or their tenure. Because my objective was to explore the institutional setting and the impacts of state–industry–community relations, I chose to anonymize my informants, only citing them as representatives of their group or institution. Some informants will be traceable to certain groups or institutions by their position, but in these cases, the informants have spoken on behalf of their respective institution/group. None of my informants specifically asked for anonymity, but several times, they avoided sensitive topics concerning agreements and tenure.

The principle of information underpins the meaning of informed consent. Informed consent entails giving as much information as possible about the research so that prospective participants can make an informed decision on their involvement. Given my thorough presentation of the project, such consent has normally been granted in that my informants have been willing to speak to me or have invited me into their gatherings and social activities. I did not obtain written consent; as such, a procedure would have hampered the necessary informality of the situation and risked introducing an awkward level of formality into the information flow.

My independence and impartiality as a researcher must be clear, and any conflicts of interest or partiality must be explicit. This was particularly important given that I engaged a wide variety of stakeholders in the contested energy spaces of Wood Buffalo. I have made an effort to portray the different interests and stakes presented to me in as thorough and correct a manner as possible, which is hopefully in line with the objectives of the stakeholders. To the best of my knowledge, I have no personal interests or affiliations that may create conflicts of interest in relation to either of my informants or their respective institutions or groups.

References

Aase, T. H., & Fossåskaret, E. (2007). *Skapte virkeligheter: kvalitativt orientert metode*. Oslo, Norway: Universitetsforlaget.

Allen, J. (2003). *Lost geographies of power*. Oxford, UK: Blackwell Publishing.

Allen, J. (2011a). Powerful assemblages? *Area, 43*(2), 154–157. https://doi.org/10.1111/j.1475-4762.2011.01005.x

Allen, J. (2011b). Topological twists: Power's shifting geographies. *Dialogues in Human Geography, 1*(3), 283–298. https://doi.org/10.1177/2043820611421546

Atkinson, P., Delamont, S., & Coffey, A. (2004). *Key themes in qualitative research: Continuities and changes*. Walnut Creek, CA: Rowman Altamira.

Becker, H. S. (2008). *Tricks of the trade: How to think about your research while you're doing it*. Chicago, IL: University of Chicago Press.

Blumer, H. (1980). Mead and blumer: The convergent methodological perspectives of social behaviorism and symbolic interactionism. *American Sociological Review, 45*, 409–419.

Bryant, A. (2009). *Grounded theory and pragmatism: The curious case of Anselm Strauss*. Paper presented at the Forum Qualitative Sozialforschung/Forum: Qualitative Social Research.

Bryman, A. (1988). In M. Bulmer (Ed.), *Quantity and quality in social research, Contemporary social research* (Vol. 18). London, UK: Routledge.

Charmaz, K. (2014). *Constructing grounded theory*. Los Angeles, CA: Sage.

Charmaz, K., & Mitchell, R. G. (1996). The myth of silent authorship: Self, substance, and style in ethnographic writing. *Symbolic Interaction, 19*(4), 285–302.

Clarke, A. (2005). *Situational analysis: Grounded theory after the postmodern turn*. Thousand Oaks, CA: Sage.

Clarke, A. E. (2003). Situational analyses: Grounded theory mapping after the postmodern turn. *Symbolic Interaction, 26*(4), 553–576.

Clarke, T. (2009). *Tar sands showdown: Canada and the new politics of oil in an age of climate change*. Toronto, Canada: J. Lorimer.

Crampton, J. W., & Elden, S. (2007). *Space, knowledge and power: Foucault and geography*. Aldershot, UK: Ashgate Publishing.

DeLanda, M. (2006). *A new philosophy of society: Assemblage theory and social complexity*. London, UK: A&C Black.

DeLanda, M. (2016). *Assemblage theory*. Edinburgh, Scotland: Edinburgh University Press.

Denzin, N., & Lincoln, Y. (2000). Introduction: The discipline and practice of qualitative research. In N. Denzin & Y. Lincolnm (Eds.), *Handbook of qualiative research*. Thousand Oaks, CA: Sage Publications.

Dewey, J. (1916). The pragmatism of Peirce. *The Journal of Philosophy, Psychology and Scientific Methods, 13*, 709–715.

Fassin, Y. (2009). The stakeholder model refined. *Journal of Business Ethics, 84*(1), 113–135.

Freeman, R. E. (2010). *Strategic management: A stakeholder approach*. New York, NY: Cambridge University Press.

Freeman, R. E., & Velamuri, S. R. (2006). A new approach to CSR: Company stakeholder responsibility. In *Corporate social responsibility* (pp. 9–23). Berlin, Germany: Springer.

Frooman, J. (1999). Stakeholder influence strategies. *Academy of Management Review, 24*(2), 191–205.

Gadamer, H.-G. (2006). Classical and philosophical hermeneutics. *Theory, Culture & Society, 23*(1), 29–56.

Glaser, B. G. (1978). *Theoretical sensitivity: Advances in the methodology of grounded theory*. Mill Valley, CA: Sociology Press.

Glaser, B. G. (2002). *Constructivist grounded theory?* Paper presented at the Forum qualitative sozialforschung/forum: Qualitative social research.

Glaser, B. G., & Strauss, A. L. (2009). *The discovery of grounded theory: Strategies for qualitative research*. Piscataway, NJ: Transaction Publishers.

Gobo, G. (2008). Re-conceptualizing generalization: Old issues in a new frame. In *The SAGE handbook of social research methods* (pp. 193–213). Los Angeles, CA: Sage.

Haarstad, H. (2009). *Changing conditions for political practice: FDI discourse and political spaces for labor in Bolivia*. Bergen, Norway: The University of Bergen.

Holt-Jensen, A. (2009). *Geography-history and concepts: A student's guide* (4th ed.). Los Angeles, CA: Sage.

Howitt, R., & Stevens, S. (2005). *Cross-cultural research: Ethics, methods and relationships*. Oxford, UK: Oxford University Press.

Huseman, J., & Short, D. (2012). 'A slow industrial genocide': Tar sands and the indigenous peoples of northern Alberta. *The International Journal of Human Rights, 16*(1), 216–237.

Jaworski, A., & Coupland, J. (2005). Othering in gossip: "You go out you have a laugh and you can pull yeah okay but like…". *Language in Society, 34*(5), 667–694.

Jones, O. (2008). Stepping from the wreckage: Geography, pragmatism and anti-representational theory. *Geoforum, 39*(4), 1600–1612. https://doi.org/10.1016/j.geoforum.2007.10.003

Lackey, J., & Sosa, E. (2006). *The epistemology of testimony*. Oxford, UK: Oxford University Press.

Le Billon, P., & Carter, A. (2012). Securing Alberta's tar sands: Resistance and criminalization on a new energy frontier. In *Natural resources and social conflict: Towards critical environmental security* (pp. 170–192). Basingstoke, UK: Palgrave Macmillan.

Mason, J. (2002). *Qualitative researching*. London, UK: Sage.

Mitchell, R. K., Agle, B. R., & Wood, D. J. (1997). Toward a theory of stakeholder identification and salience: Defining the principle of who and what really counts. *Academy of Management Review, 22*(4), 853–886.

Mitchell, T. (2002). *Rule of experts: Egypt, techno-politics, modernity*. Berkeley, CA: University of California Press.

Næss, H. E. (2015). *Globografi. En kort innføring i flerlokalitetsforskning*. Oslo, Norway: Cappelen Damm.

Nelson, C., Treichler, P. A., & Grossberg, L. (1992). Cultural studies: An introduction. *Cultural Studies, 1*, 22.

Nikiforuk, A. (2010). *Tar sands: Dirty oil and the future of a continent*. Vancouver, Canada: Greystone Books.

Reichertz, J. (2007). *Abduction: The logic of discovery of grounded theory*. London, UK: Sage.

Rennie, D. L. (2000). Grounded theory methodology as methodical hermeneutics: Reconciling realism and relativism. *Theory & Psychology, 10*(4), 481–502.

Robbins, P., & Marks, B. (2010). Assemblage geographies. In *The Sage handbook of social geographies* (p. 176). Los Angeles, CA: Sage.

Rorty, R. (1999). *Philosophy and social hope*. London, UK: Penguin.

Silverman, D. (2013). *Doing qualitative research: A practical handbook*. Los Angeles, CA: SAGE Publications.

Steup, M., Turri, J., & Sosa, E. (2013). *Contemporary debates in epistemology*. Hoboken, NJ: Wiley.

Therborn, G. (2011). *The world: A beginner's guide*. Cambridge, UK: Polity.

Wadel, C. (1991). *Feltarbeid i egen kultur: en innføring i kvalitativt orientert samfunnsforskning*. Flekkefjord, Norway: Seek.

Wanvik, T. I. (2015). Globografi - En kort innføring i flerlokalitetsforskning av Hans Erik Næss. *Norsk Geografisk Tidsskrift, 69*, 312–313.

Xiang, B. (2013). Multi-scalar ethnography: An approach for critical engagement with migration and social change. *Ethnography, 14*(3), 282–299.

Yin, R. K. (2013). *Case study research: Design and methods*. London, UK: Sage Publications.

Chapter 4
Zooming in on Contested Energy Spaces: The Study Area

Abstract This chapter gives an account of the geographical features of the contested energy spaces of the Canadian North, and their relationship to the wider Canadian federation. A selection of communities has been chosen to represent a variety of responses to the massive industrial developments in the Regional Municipality of Wood Buffalo, Alberta. It also gives a brief summary of the Indigenous Métis history, and how certain features of these Indigenous communities could be said to be particularly adaptive and pragmatically oriented towards rupture and change.

Keywords Canada · Alberta · Wood Buffalo · Conklin · McKay · McMurray · Métis · Indigenous

To a Norwegian, Canada at first glance may be considered to be a part of the family. Our people share many of the same cultural traits, values, habits and institutional robustness. However, Canada is so much more (Fig. 4.1).

From the wild, moist, mountainous west coast of British Columbia to the cold and vast Northern Territories, and to the seemingly empty, never-ending plains of the Prairie Provinces, the country is wildly varied and complex. Canada is sparsely populated; the majority of its land territory is dominated by forest and tundra and the Rocky Mountains. However, it is highly urbanized, with 82% of the 35.15 million people concentrated in large- and medium-sized cities, many near the southern border.

Since the early twentieth century, the growth of Canada's manufacturing, mining and service sectors has transformed the nation from a largely rural economy to an urbanized, industrial one. Like many other developed nations, the Canadian economy is dominated by the service industry, which employs about three-quarters of the country's workforce (Statistics Canada, 2017). However, Canada is unusual among developed countries in the importance of its primary sector, of which the forestry and petroleum industries are two of the most prominent components. In Alberta, home of the largest oil extraction operation in the northern hemisphere, and thus the focal point of this research, the numbers are even more skewed.

T. I. Wanvik, *Contested Energy Spaces*, SpringerBriefs in Geography, https://doi.org/10.1007/978-3-030-02396-6_4

Fig. 4.1 Canada is so much more. The Rocky Mountains and the Alberta Prairie. From first field trip, 2014. Photos by Laura M. I. Wanvik

Alberta, the westernmost of Canada's three Prairie Provinces, shares many physical features with its neighbours to the east: Saskatchewan and Manitoba. The Rocky Mountains form the southern portion of Alberta's western boundary with British Columbia. Alberta was named after Princess Louise Caroline Alberta, fourth daughter of Queen Victoria. The province is home to the country's largest deposits of oil and natural gas. The oil sands have the third largest oil reserves in the world, after Venezuela and Saudi Arabia.

With 11% of the total population of Canada, Alberta accounts for around 17% of its gross domestic product (GDP), 28% of which is derived directly from the energy sector (Alberta Energy, 2017). In 2014, the oil and gas industry produced one-quarter of Alberta's GDP, almost 70% of its exports and 35% of Alberta government's revenues, and the industry accounted for just under 150,000 direct and indirect jobs (Alberta Energy, 2017). However, these numbers do not account for important factors such as the volatility of the unconventional oil industry: the combination of fiscal and taxation policy that creates significant concerns related to revenue realization, the weakening of various regulatory regimes under pressure from the short-term priorities of the oil industry and the costs of various externalities, such as pollution and inflation (Shrivastava & Stefanick, 2015). Government policy responses to these challenges have long followed a familiar trajectory by ceding more regulatory control to industry, opening new doors for foreign acquisitions and raising financial, social and environmental subsidies (Shrivastava & Stefanick, 2015; Wanvik, 2016).

Albertans have always had "a mind of their own", distancing themselves from the federal government in Ottawa and to a large extent from the rest of Canada. The province's first decade was prosperous; immigration accelerated, grain harvests were bountiful, new communities settled, and its network of railway lines expanded rapidly. Yet, resentment grew among farmers, who believed that the railways, banks and grain-elevator companies were jeopardizing their status as independent entrepreneurs (Canadian Encyclopedia, 2017; Wanvik, 2016). Alberta's dissent from the policies of the federal government has continued to this day.

One of the reasons for this dissent has been a particular characteristic of the Albertan economy. For nearly a century, Alberta's economy has relied on primary resource exploitation and subsequent dependence on foreign markets, moving from the export of fur, prior to becoming a province, to wheat and beef, and finally to petroleum. Although the existence of vast bitumen deposits has been known for decades, it was only with advances in extraction technologies and a rapid rise in international oil prices during the last part of the twentieth century that the production of unconventional oil became profitable.

The rapid expansion of the oil sector has been achieved with significant government support for Alberta's oil industry in the form of investment, subsidies and tax breaks at both the federal and provincial levels (Shrivastava & Stefanick, 2015). Alberta's extremely industry-friendly tax and revenue sharing regime, along with the province's propensity to externalize the social and environmental costs of bitumen oil production, has led to handsome returns for private corporations (Campanella, 2012).

The growing economic and political might of Alberta has made this province the barometer of political economic change in Canada. The rising political influence of this landlocked province can also be construed as leading to the "Albertization" of Canada under the previous federal government (2006–2015). This political orientation includes government austerity, especially with respect to social programmes, privatization of government services and a reduction in income tax for corporations and upper-income earners (Shrivastava & Stefanick, 2015).

Alberta is home to a large proportion of the indigenous communities inhabiting the North American continent. Nearly, one in six indigenous peoples in Canada lives in the Prairie Province, numbering approximately 221,000. They make up 6% of the total population of the province. Nearly, half of these are Métis, descendants of First Nations and European settlers, with their own unique cultural heritage (Sealey & Lussier, 1975), recognized as Aboriginal peoples in 1982 (Pulla, 2013) and granted "Indian" status as of 2016 (SCC, 2016).

4.1   The Métis

The Métis have historically been relegated to the status of "Canada's forgotten people" (Lischke & McNab, 2007; Sealey & Lussier, 1975). Métis communities are typically ignored in treaty negotiations and land claims agreements, because they have no recognized land base (with the exception of the Alberta Settlements). Without recognition of their land title, the ability of the Métis to raise revenues from specific land or resource development projects or to negotiate directly with land and business developers remains a challenge (Dubois & Saunders, 2013; Madden, 2008; Weinstein, 2007). With the very existence of their collective title to land denied by the federal government, the Métis must instead find other ways of asserting their right to recognition and self-determination. With their attitude of "just do it", the Métis seek innovative ways of achieving self-government (Pulla, 2013), by bringing their claims through the courts (Weinstein, 2007) or through consultation processes and IBA negotiations (Wanvik, 2016).

For centuries, marginalized Métis communities in Wood Buffalo, Alberta and in Canada in general have struggled for recognition as rights-bearing communities. Generations have fought an uphill battle against prospectors, governments and industry—recently exemplified by the all-encompassing industrial adventures of the oil sands—to safeguard their rightful share of the riches from their traditional territories.

Some have claimed that "the very concept of Métis, as a people, challenged the established boundaries of culture in Canada" (Teillet, 2013, p. 7). Unlike the First Nations and the Inuit, the federal government has only recently recognized the Métis as Indians under section 91(24) of the Constitution Act 1867, so their well-being has not been formally recognized as a federal responsibility. After the Supreme Court case of *Daniels* versus *Canada*, the Crown received a certain fiduciary responsibility for the Métis (SCC, 2016). It was stated that Parliament has legislative authority for all Indigenous peoples, and it is the federal government to which the Métis and non-status Indians can turn. This simple clarification unblocks the federal government's self-created obstacle to negotiations with these Indigenous groups (Madden, 2016).

That the Métis share a sense of nationhood and collective consciousness as a distinct Indigenous people has been well documented (See for example Adams, Dahl, & Peach, 2013; Andersen, 2008; Chartier, 1994; Chartrand, 2008; Madden, 2008; Sawchuk, 1985; Teillet, 2007, 2013; Weinstein, 2007). Métis marginalization can be attributed to several factors, notably their unique history and differences in legal and policy positions of the federal government towards Indigenous peoples in Canada (Dubois & Saunders, 2013). In the words of Métis leader Clément Chartier, "being Métis is more than being of mixed blood: there is language, heritage and a way of life" (Chartier, 1994, p. 82).

The Michif concept of *kaa-tipeyimishoyaahk* implies the Métis notion of being "people who own themselves", implying an embodied understanding of independence and self-sufficiency (Gaudry, 2014). While formal constitutional recognition of their inherent right to self-government (among other defined rights) remains a goal, the Métis in the face of these challenges have had to find innovative ways to pursue self-government initiatives on their own (Madden, Graham, & Wilson, 2005; Teillet, 2007, 2013).

4.2   Wood Buffalo and Local Communities

Oil began forming in southern Alberta when tiny marine creatures died and drifted to the sea-floor in prehistoric times. Over time, their bodies were compressed by heat and pressure and formed liquid rock oil—referred to today as petroleum. In the north, rivers flowing away from the sea deposited sand and sediment. When tectonic plates shifted to form the Rocky Mountains, the pressure squeezed the oil northwards, causing it to seep into the sand (Canada Oil Sands, 2017).

The Regional Municipality of Wood Buffalo (RMWB) is a specialized municipality located in north-eastern Alberta. Formed as a result of the amalgamation of the City of Fort McMurray and Improvement District No. 143 in 1995, it is the second largest municipality in Alberta in terms of area. It is home to vast oil sand deposits, also known as the Athabasca Oil Sands, making the region one of the fastest growing industrial areas in Canada. The oil sands region of Canada is primarily situated in the north-eastern part of the province of Alberta, until recently relatively sparsely populated by various Indigenous groups of First Nations or Métis origin. As in many contested energy spaces, the oil sands deposits are found beneath the traditional lands of Indigenous people. One of the fastest growing Indigenous groups in Alberta is the Métis, with its close to 100,000 members, representing more than 21.4% of all the Métis of Canada (Alberta Métis, 2017) (Fig. 4.2).[1]

According to the 2006 census data, the Indigenous Métis population accounted for 4.9% of the total population, while the First Nations population was 4.7%, indicating around 2500 people in each group. Indigenous groups constitute the majority of the rural population of Wood Buffalo, where the total number of inhabitants is 3500. The shadow population[2] of the rural areas outnumbered the locals by a factor of 10 (38,000) in 2015 (Government of Canada, 2015).[3]

The three local communities of interest in this study are among the dominant Métis communities in the region and were chosen because of their location in the midst of the oil sands projects. Fort McMurray Métis is an urban Métis community in the regional urban centre of Fort McMurray. McKay Métis is found between the massive industrial open mining pits north of the urban centre. Conklin is situated

---

[1] http://albertametis.com/about/.

[2] "Shadow population" refers to temporary residents of a municipality who are employed by an industrial or commercial establishment in the municipality for a minimum of 30 days within a municipal census year.

[3] http://municipalaffairs.alberta.ca/documents/msb/2015_Municipal_Affairs_Population_List.pdf.

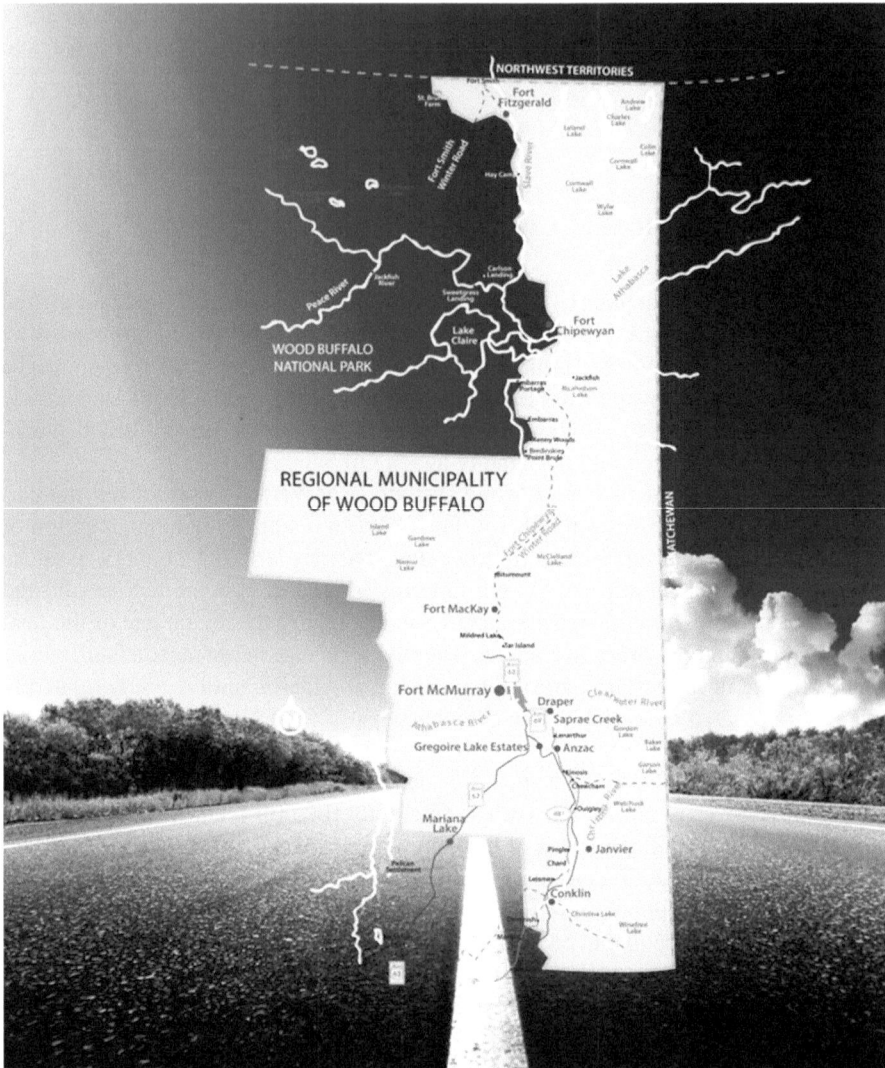

Fig. 4.2 Regional municipality of Wood Buffalo. Source: Fort McMurray Tourism 2017

further south, by Christina Lake along highway 881, in an area dominated by steam-assisted gravity drainage production (for reflections on sampling, see Sect. 3.6).

My preliminary findings all showed some positive bias, albeit hesitant and conditional, towards industrial development of their traditional territories, and they all related to and negotiated with industrial players in their vicinity in different ways.

4.3 Fort McMurray

The first community considered in this research is McMurray Métis, situated in and around the boom town of Fort McMurray. Nestled in a forest valley where the Athabasca and Clearwater rivers meet is Wood Buffalo's urban centre, the community of Fort McMurray, with around 80,000 inhabitants. Originally established as a Hudson's Bay Company trading post in 1870, today Fort McMurray attracts attention from around the world as the residential and commercial focal point of Canada's oil sands industry. The community has played a significant role in the history of the petroleum industry in Canada. Oil exploration is known to have occurred in the early twentieth century, but Fort McMurray's population remained small, with no more than a few hundred people. In 1967, the Great Canadian Oil Sands (now Suncor) plant opened, and Fort McMurray's growth soon took off. On April 1, 1995, the City of Fort McMurray and Improvement District No. 143 were amalgamated to form the Regional Municipality of Wood Buffalo (Government of Canada, 2017).[4] As a result, Fort McMurray was no longer officially designated a city. Instead, it was designated an urban service area within a specialized municipality. The amalgamation placed the entire regional municipality of Wood Buffalo under a single government. Its municipal office is located in Fort McMurray.

McMurray Métis was founded in 1987, and it is governed under the by-laws of the Métis Nation of Alberta. If the province of Alberta recognizes the McMurray Métis as a historical community, then the 300 members of the locality not only will be eligible for the same hunting and trapping rights that other Métis communities enjoy but also will make it mandatory for oil and gas companies to consult them (McDermott, 2015). The McMurray Métis community represents a subregional hub within the wider Lac La Biche regional Métis community that extends from Île-à-la-Crosse, Saskatchewan to Fort McMurray and down to Beaver Lake and Lac La Biche (Clark, O'Connor, & Fortna, 2015). The centrality of the waterways to Fort McMurray has led to a strong Métis presence. The river system was the initial means of transport connecting the north-east of Alberta with the "historic trail" to the west through Saskatchewan, and the Métis played a crucial role in the navigation of the rivers, from scows to steamships. Oral history accounts are clear that the House River–McMurray–La Loche axis along the Clearwater was heavily Métis (Clark et al., 2015).

4.4 Fort McKay

The second community, Fort McKay Métis, is located approximately 45 km north of Fort McMurray, in the centre of most oil sands mining operations in north-eastern Alberta. Founded by the Métis during Canada's fur trade in the early 1800s, it is

---

[4] http://www.municipalaffairs.alberta.ca/cfml/MunicipalProfiles/index.cfm?fuseaction=BasicRep ort&MunicipalityType=SMUN&stakeholder=508&profileType=HIST.

now home to Cree, Dene and Métis residents. The Fort McKay Métis community is historically and contemporarily connected to the larger Métis Nation, recognized in s. 35 of the Constitution as one of the Aboriginal peoples of Canada with distinct rights. The Hudson's Bay Company established Fort McKay in 1820. Until the oil companies arrived in earnest 150 years later, its people led a largely hunting-and-trapping existence (McCarthy, 2015).[5]

The Fort McKay Métis Community is made up of the historic Métis community that originally provided labour for the fur trade in the Athabasca region of what is now north-eastern Alberta in the early nineteenth century. Its members have a mixed ancestry that includes French, English, Cree, Dene and Métis heritage with close ties to members of the Fort McKay First Nation (Fort McKay Métis, 2017). Located in the heart of the oil sands, the tiny community of between 44 and 300 inhabitants[6] has faced unprecedented change over the past 30 years. This change has brought both opportunities and challenges, but by following the elders' traditional teachings and committing to grass-roots development, the community is facing its challenges head on while taking advantage of the opportunities presented (Fort McKay Métis, 2017).

4.5   Conklin

The third community is Conklin Métis. Despite its isolation, Conklin is at the centre of the oil sands development. Here, the Métis people have practised trapping, hunting, fishing and harvesting for over 100 years, living off the land. Steam-assisted gravity drainage operations combined with ancillary high-voltage transmission lines and bitumen pipelines have had a great impact on the area. There are currently 337 people living in Conklin, according to the latest census.[7] According to the community, the traditional harvesting territory of the Conklin Métis covers about 10,000 km^2, stretching from the Wiau and Grist lakes in the south to the Algar and Gordon lakes in the north (Golder Associates, 2011).

According to the traditional land use studies from the area, the traditional way of life, based on hunting, fishing, trapping and gathering, is quickly becoming impossible for the Métis of Conklin (Conklin Métis Local #193, 2012). Today, community members claim to find it increasingly difficult to access traditional lands. Old trails have been destroyed or upgraded into roads for trucking, numerous new seismic cut lines have been created throughout formerly intact lands and long-standing routes have been restricted or blocked by oil developers (Conklin Métis Local #193, 2012, p. 36). Development has caused a rapid decline in the numbers of ani-

---

[5] http://www.theglobeandmail.com/news/alberta/where-oil-and-water-mix-oil-sands-development-leaves-fort-mckays-indigenous-communitytorn/article27151333/.

[6] The numbers are contested, ranging from 44 recorded inhabitants of the hamlet of Fort McKay to the "more than 300" claimed by local leaders (interview 2015).

[7] http://www.rmwb.ca/living/Communities/Conklin.htm.

mals, berries and plants, as well as a decrease in air and water quality (ibid: 72). In addition, the social and cultural challenges experienced within the small community are devastating, with substance abuse, alcoholism, high crime rates and poor living conditions taking a heavy toll of its inhabitants (my own field work). Lately, the homelessness situation in Conklin has reached a crisis point, and in a recent report on the issue say historic failures, large-scale environmental manipulation and bureaucratic red tape is to blame for this hardship (Fortna, 2018). The failure particularly by the government to take action and provide land and housing in Conklin is having a catastrophic effect on the community, particularly among the young adults who did not receive the opportunity to obtain land through the land tenure program. These young men and women are now being forced to live with their parents and grandparents in contemporary housing.

The history of the Conklin Métis is a microcosm of the complex history of the Métis (and other Indigenous peoples) in Canada, which has been characterized throughout modern Canadian history by recurring cycles of settlement, displacement, dispossession and dispersion of Métis people from traditional homelands and movements to new lands (Conklin Métis Local #193, 2012). The province of Alberta and the regional municipality of Wood Buffalo struggle to govern this complex landscape of industrial developments and traditional ways of life.

References

Adams, C., Dahl, G., & Peach, I. (2013). *Métis in Canada: History, identity, law and politics.* Edmonton, Canada: University of Alberta.

Alberta Energy. (2017). Resource revenues collected. Government of Alberta. Accessed October 20. http://www.energy.alberta.ca/About_Us/2564.asp.

Alberta Métis. (2017). Métis Nation. Accessed 17 Oct 2018. http://albertametis.com/.

Andersen, C. (2008). From nation to population: The racialisation of 'Métis' in the Canadian census. *Nations and Nationalism, 14*(2), 347–368.

Campanella, D. (2012). *Misplaced generosity: Update 2012: Extraordinary profits in Alberta's oil and gas industry.* Edmonton, Canada: Parkland Institute.

Canada Oil Sands. (2017). *Oil sands history and milestones.* Retrieved from http://www.canada-soilsands.ca/en/what-are-the-oil-sands/oil-sands-history-and-milestones

Canadian Encyclopedia. (2017). *Alberta.* Retrieved from http://www.thecanadianencyclopedia.ca/en/article/alberta/

Chartier, C. (1994). Métis perspective: Continuing Poundmaker and Riel's quest. In Royal commission on Aboriginal Peoples (Ed.), *Perspectives and realities.* Ottawa, Canada: Government of Canada.

Chartrand, L. (2008). 'We rise again': Metis traditional governance and the claim to Metis self-government. In Y. Belanger (Ed.), *Aboriginal self-government in Canada* (3rd ed., pp. 145–157). Saskatoon, Canada: Purich Publishing.

Clark, T. D., O'Connor, D., & Fortna, P. (2015). *Mark of the Metis - Fort McMurray: Historic and contemporary rights-bearing Métis community.* Retrieved from Alberta, Canada.

Conklin Métis Local #193. (2012). *Stories from a long time ago.* Retrieved from Conklin, Alberta, Canada.

Dubois, J., & Saunders, K. (2013). "Just do it!": Carving out a space for the Métis in Canadian Federalism. *Canadian Journal of Political Science, 46*(1), 187–214.

Fort McKay Métis. (2018). History of the McKay Métis. Fort McKay Métis. Accessed 29.08. http://www.fortmckaymetis.com/history.html.

Fortna, P. (2018). *How much longer: A preliminary assessment of homelessness in Conklin.* Retrieved from Fort McMurray.

Gaudry, A. J. P. (2014). *Kaa-tipeyimishoyaahk - 'We are those who own ourselves': A political history of Métis self-determination in the north-west, 1830-1870* [PhD Monograph]. Victoria, Canada: University of Victoria.

Golder Associates. (2011). *Leismer to Kettle River Crossover Project* (1013340049/6000/6001). Retrieved from Calgary.

Government of Canada. (2015). *Growing trade and expanding msarkets.* edited by Finance. Ottawa: Government of Canada.

Lischke, U., & McNab, D. T. (2007). *The long journey of a forgotten people: Métis identities and family histories.* Waterloo, Canada: Wilfrid Laurier University Press.

Madden, J. (2008). The Métis nation's self-government agenda: Issues and options for the future. In *Métis-crown relations: Rights, identity, jurisdiction, and governance* (pp. 233–290). Toronto, Canada: Irwin Law.

Madden, J. (2016, April 18). Daniels v. Canada: A case of simple answers with significant consequences. *Candian Lawyer.*

Madden, J., Graham, J., & Wilson, J. (2005). *Exploring options for Métis governance in the 21st century.* Ottawa, Canada: Institute on Governance Ottawa.

McCarthy, S. (2015, February 17). 'Anti-petroleum' movement a growing security threat to Canada, RCMP say, Politics. *Globe and Mail.* Retrieved from http://www.theglobeandmail.com/news/politics/anti-petroleum-movement-a-growing-security-threat-to-canada-rcmp-say/article23019252/

McDermott, V. (2015, August 17). McMurray Métis passes as a historical, rights-based community, argues report. *Fort McMurray Today.* Retrieved from http://www.fortmcmurraytoday.com/2015/08/17/report-arguing-mcmurray-metis-passes-as-a-historical-community-would-make-industry-consultation-mandatory

Pulla, S. P. (2013). Regional nationalism or national mobilization? A brief social history of the development of Métis political organization in Canada, 1815-2011. *Metis in Canada, 397*-431.

Sawchuk, J. (1985). The Metis, non-status Indians and the new Aboriginality: Government influence on native political alliances and identity. *Canadian Ethnic Studies= Etudes Ethniques au Canada, 17*(2), 135.

SCC. (2016). Daniels v. Canada (Indian Affairs and Northern Development). Canada: Supreme Court of Canada

Sealey, D. B., & Lussier, A. S. (1975). *The Metis: Canada's forgotten people.* Winnipeg: Pemmican Publications.

Shrivastava, M., & Stefanick, L. (2015). *Alberta oil and the decline of democracy in Canada.* Edmonton, Canada: Athabasca University Press.

Statistics Canada. (2017). Canada at a glance. Government of Canada. Accessed 17 Oct 2018. https://www150.statcan.gc.ca/n1/pub/12-581-x/2017000/pop-eng.htm.

Teillet, J. (2007). The winds of change: Métis rights after Powley, Taku, and Haida. In *The long journey of a forgotten people: Métis identities and histories* (pp. 57–78). Waterloo, Canada: Wilfrid Laurier University Press.

Teillet, J. (2013). *Métis law in Canada* (0991702700). Retrieved from Alberta, Canada. http://albertametis.com/wp-content/uploads/2014/04/Metis-Law-in-Canada-2013-1.pdf

Wanvik, T. I. (2016). Governance transformed into CSR – new governance innovations in the Canadian oil sands. *The Extractive Industries and Society, 3*, 517.

Weinstein, J. (2007). *Quiet revolution west: The rebirth of Métis nationalism.* Calgary, Canada: Fifth House.

Chapter 5
Challenging the Permanence of Contested Energy Spaces

Abstract In this chapter, the author argues that the stability and permanence of society's relationship with energy tends to be overstressed. While there are obvious structures of inertia and permanence, the energy–society nexus is also characterized by rupture, unpredictability and instability. It is worth debating whether the prevalent vocabulary of "regimes" and "lock-in" has certain incapacitating consequences and whether we should pay more attention to volatility and change. By employing assemblage thinking, this chapter investigates the instabilities of landscapes, governance and power within contested energy spaces.

Keywords Energy · Instability · Carbonscapes · Assemblage · Transformation

So far, I have elaborated on the conceptual foundations of contested energy spaces. In this section, I underscore the instability of energy landscapes (Haarstad & Wanvik, 2016). Several scholars have noted recent re-emergence of energy as a concern for geographers (Bridge, 2017; Bridge, Bouzarovski, Bradshaw, & Eyre, 2013; Calvert, 2016; Pasqualetti & Brown, 2014; Zimmerer, 2011). There also appears to be corresponding trends in related fields outside of geography, such as anthropology (Boyer, 2011), sociology and critical theory (see special issue of *Theory, Culture, Society*, 2014, vol. 5, issue 3) and history (Kander, Malanima, & Warde, 2014). Even though "energy geography" can be considered a distinct sub-field of geography (Calvert, 2016), it is informed by a wide range of scholarship. Energy is also a key topic of enquiry in studies of natural resources, political economy, cities and other interrelated fields (Bakker & Bridge, 2006; Calvert, 2016).

According to recent geographical and social science scholarship (i.e. Huber, 2013; Urry, 2013, 2014; Watts, 2013), the political and material landscapes of our fossil fuel society are as robust as ever. The interweaving of material, social and cultural forms and artefacts creates solid structures of "petroculture" (Marriott & Minio-Paluello, 2012), "fossil capitalism" (Huber, 2013; Watts, 2013), "carbon lock-in" (Unruh, 2000) or "carbon democracy" (Mitchell, 2011). Urry (2014, p. 3) states that energy systems and their lock-ins are "not subject to simple human intervention and control". In much of the geographical and social science scholarship, there is a tendency to stress path dependencies and inertia that shape society's relationships with energy (Bridge et al., 2013; Bulkeley, Castán Broto, & Maassen, 2013, 2014; Calvert, 2016; Hodson & Marvin, 2010; Rutherford & Coutard, 2014; Shove, Watson, & Spurling, 2015). Calvert (2016) suggests that in recent revitalization of energy geographies, there has been a greater stress on the political, economic, technological and cultural work done to establish and maintain energy systems.

However, I argue that the stability and permanence of society's relationship with carbon tends to be overstressed. While there are obvious structures of inertia and permanence, the carbon–society nexus is also characterized by rupture, unpredictability and instability. It is worth debating whether the prevalent vocabulary of "regimes" and "lock-in" has certain incapacitating consequences and whether we should pay more attention to volatility and change. If not, I would suggest that there is a danger that we mimic the narrative of the inescapability of oil that the fossil fuel industry has carefully constructed. We will also miss important opportunities to hone conceptual frameworks in energy geography.

One of the main aims of this book to is to reconceptualize the socio-material landscapes created by fossil-based energy systems. I suggest that we need to appreciate their instabilities and identifying windows for transformation. Contested energy spaces based on oil, what I elsewhere have termed Carbonscapes (Haarstad & Wanvik, 2016), then, are the spaces created by material expressions of carbon-based energy systems and the institutional and cultural practices attached to them. While a common theoretical stance is to depict the co-articulation of these elements as a coherent totality or as a stable organic whole, I want to theorize carbonscapes as more contingent. I suggest that assemblage thinking, which is gaining ground in geography, provides a set of conceptual tools with which to identify these contingencies. Assemblage thinking promotes an ontology that dismisses the idea of systems as stable, organic wholes in favour of an ontology of entities without essence that are held together in more or less impermanent relationships.

Regarding carbonscapes and contested energy spaces in general, many geographical and critical analyses have been developed as a critique of the mainstream and hegemonic "resource curse" literature. Literature on "the curse", dominated by economics and political science, has seen energy spaces as cursed by economic and political processes at the national scale (see, for example, Humphreys, Sachs, & Stiglitz, 2007; Mehlum, Moene, & Torvik, 2006). Geographers, anthropologists and others have argued that the discontents of many energy spaces are far more complex and must be understood in terms of both skewed distribution of costs and benefits locally, enclave formation and spaces of enclosure, and unequal integration with the global political economy of oil (Bebbington, Hinojosa, Bebbington, Burneo, & Warnaars, 2008; Haarstad, 2014b; Haarstad & Wanvik, 2016; Kirshner & Power, 2015; Logan & McNeish, 2012; Stevens & Dietsche, 2008). Yet in broadening out the scope and complexity of the processes underlying "the curse", geographers tend to deepen the view of the grip of that oil extraction has on social development trajectories. Watts (2004), for example, has suggested that we should be attentive to how oil is "inserted into an already existing political landscape of forces, identities and forms of power" (2004, p. 76). Elsewhere, he draws attention to the global regime of accumulation that envelops oil extraction (Watts, 2013). The operative perspective in this literature is typically that the local energy spaces (and their patterns of underdevelopment, inequalities and environmental disruption) are tortuously embedded in the broader political economy: multiscalar complexes involving oil companies, political institutions and more. Much of this work is rather persuasive, foregrounding the power structures of the global regime of oil as an important driver within this rationale. Yet in this sense, it tends to present an image of relative stability and resistance to change, as local dynamics are closely embedded within the globalized regime. Even though the contentious politics of social movements and civil society always remain part of the picture (see Bebbington et al., 2008; Haarstad, 2012; Haarstad & Fløysand, 2007; Perreault, 2006), the general perspective seems to be that the hegemony of oil capital is able to subvert serious challenges to continued accumulation.

5.1 What Is Socio-Technical Change?

Academic discourse on changes in society's relationship with energy and technology revolves around the socio-technical transitions literature and the seminal contribution by Rip and Kemp (Rip & Kemp, 1998), which takes as a starting point that established technologies are highly intertwined with "technological regimes", i.e. the rule-set embedded in practices, skills and procedures that arbitrate how specific technologies are implemented in society), and the larger social, economic and political "socio-technical landscape" in which technological innovations arrive. A key idea is that opportunities for change are fostered in protected niches, and that actual change depends on how these niches interact with broader regimes and landscapes (Kemp, Schot, & Hoogma, 1998). This basic framework has evolved into different strands—transition management, strategic niche management, the multi-level perspective on sustainability transitions and technological innovation systems (Markard, Raven, & Truffler, 2012)—that each conceptualize relationships between stability and change in different ways.

Geographers have critiqued the under-theorized and tenuous spatial assumptions of the multi-level perspective (Coenen, Benneworth, & Truffler, 2012; Hansen & Coenen, 2015) but have also suggested ways to employ spatial vocabularies to inform perspectives within the socio-technical transitions discourse (Bridge et al., 2013). Rather than understanding radical transformations as developing in protected niches, geographers are characteristically more oriented towards cross-spatial and multiscalar networks in which radical and transformative practices are produced (Bulkeley & Betsill, 2013; Haarstad, 2014a). Yet, the perspective has also been used productively in geography to theorize urban change and transformation (Haarstad & Wanvik, 2016). Bulkeley and co-authors (2013, 2014), for example, hold that analyses of socio-technical regime change must be understood in relation to the broader political economy of relations that go into maintaining and contesting urban infrastructures.

Others have combined various metabolic and infrastructural perspectives to examine urban socio-technical regimes and how they are contested and reconfigured (Hommels, 2005; McFarlane & Rutherford, 2008; Pflieger, Kaufmann, Pattaroni, & Christophe, 2009).

Many scholars are now using variations of the socio-technical transitions literature as an inroad to analysing processes of change due to the multidimensional conceptual framework it offers. It also has some family ties to assemblage thinking, in the sense that ideas around the social construction of technology can in part be traced to Latour and actor-network theory (see Rip & Kemp, 1998), which has a parallel inspirational legacy as assemblage thinking (Müller, 2015). In this book, I advocate for assemblage thinking framework rather than socio-technical transitions theory for several reasons. First, transitions theory is oriented primarily towards incremental changes that lead to systemic transitions over long time (Markard et al. (2012) state that transitions typically take 50 years or more), which overlooks the self-significance of pockets of radical transformation. Also, transitions theory is committed to a systems perspective, which assemblage thinking attempts to break with, because in systems thinking change only becomes significant once it affects all the other elements in the system, and systems are to a certain extent self-sustaining. Finally, an overemphasizing on political economy in human geography has typically stressed how processes of capital accumulation shape socio-spatial change (Harvey, 1989). In particular, its systemic orientation prioritizes broad and long-term changes rather than specific ruptures and instabilities in cities and elsewhere. When material infrastructures, socio-cultural artefacts and political structures are all understood as mutually reinforcing forces of conservation, the opportunities for change are difficult to identify, appreciate and theorize. In most of the literature discussed above, a key theoretical objective has been to explain permanence and fixity rather than identifying the points of leverage for change.

Contested energy spaces are characterized both by path dependencies and by rupture. Keeling and Sandlos' blog on abandoned mines in Northern Canada is a powerful example of this instability of energy spaces (Keeling & Sandlos, 2015; Sandlos & Keeling, 2013). Everything seems stable, until it suddenly does not. Energy markets themselves are subject to sudden jolts from unpredictable factors such as oil price instability, geopolitical threats to energy security and terrorism. There is a danger that the resistant nature of the landscapes made by energy is exaggerated and that the theoretical frameworks available are so populated with concepts stressing inertia that instances of change are invisible (Haarstad & Wanvik, 2016). This is a theoretical problem in the sense that it makes us unable to theorize the relationships between stability and change properly. It is also a normative problem in the sense that we as theorists may reproduce the narrative of the inevitability of certain types of energy (like oil) that the energy industry itself has so painstakingly created. The theoretical project should instead be to conceptualize contested energy spaces in ways that take account of how structures of stability coexist and are interrelated with processes of change. Our intentions here resonate with J.K. Gibson-Graham's (2006) project of destabilizing imaginaries of capitalism in ways that open spaces for negotiation and contestation.

5.2   Assemblage Thinking in Energy Geography

Assemblage thinking has recently gained ground in geography probably because it allows for the conceptualization of the entanglements of material, social and ideational elements (McFarlane & Anderson, 2011). Geographers working on energy and natural resources often use the term "assemblage" casually (Haarstad & Wanvik, 2016), as Bridge and co-authors have done when stating that "landscape describes the assemblage of natural and cultural features across a broad space and the history of their production and interaction" (2013, p. 5). Watts (2013) on his part focuses on what he terms the oil and gas assemblage: a global production network with particular properties, actors, networks, governance structures, institutions and organizations, but also a complex regime of accumulation and a mode of regulation, held together by the massive global oil infrastructure.

In a more specific theoretical sense, "assemblage" is employed to describe constellations of social and material, expressive and physical components (Allen, 2011; McFarlane & Anderson, 2011; Ogden et al., 2013; Tsing, 2005). To Sassen and Ong (2014, p. 19), the "notion of assemblage is something that helps (…) to understand transformations and perhaps even historical turning points" and is a perspective that allows us to actively destabilize powerful social categories. Assemblage thinking has together with actor-network theory been in the forefront of a revalorization of the material, or the interconnectedness between humans and non-humans (Martinez, 2007; Müller, 2015). But, even though the material realm is often associated with structure and inertia, the time–space of assemblages is imagined as inherently unstable and infused with movement and change (Marcus & Saka, 2006). As contribution to geographical thought, assemblage thinking can be useful for integrating materiality, power and scale into one single analytical framework (Wanvik, 2014).

5.3   Assemblages as Constant Emergence

Assemblages can be understood as entities without essence. They involve relations between both human and non-human components, and relational work is necessary to keep these components together. The component parts are harbouring unexercised capacities that might produce very different properties if the entities were to enter into relations with other entities. DeLanda (2006, pp. 10–11) refers to these relations as "relations of exteriority". He argues that that we must not confuse the properties of a particular entity with the capacities of its component parts to form relations with other entities. Instead of seeing social entities as totalities (organic wholes bound together by internal relations), assemblage theory stresses how the interactions between seemingly separate elements produce unstable and contingent entities, revealing the empirical stability of carbonscapes as temporary, contingent achievements, always vulnerable reconfigurations. Anderson and co-authors (2012) argue that this notion of relations of exteriority allows us to actualize ongoing processes of composition of different component parts, rethink social formations as complex wholes composed through diversity and attend to the expressive powers of entities (Bennett, 2005; Gidwani, 2008; Latour, 2005; Ong, 2007). Thus, the assemblage's only unity is that of co-functioning: it is a symbiosis, a "sympathy" (Deleuze & Parnet, 2007 [1977], p. 52). Rather than conceptualize assemblages as seamless wholes, "relations of exteriority" implies certain autonomy for the elements they relate (DeLanda, 2006, pp. 10–11).

DeLanda characterizes socio-material assemblages along three dimensions. First, he distinguishes between processes that stabilize the emergent identity of assemblages (by sharpening their borders, for example or homogenizing their composition) from those that tend to destabilize this identity and hence open the assemblage to change. These are processes of *territorialization* and *deterritorialization*, respectively (DeLanda, 2006; Deleuze & Guattari, 1988). Second, the component parts are recognized by their *emergent capacities*, properties that are contingent by the interaction with other component parts. Finally, by employing DeLanda's notion of the *assemblage converter* we can highlight the catalytic impact of well-placed component parts in either transforming assemblages or ensuring that relations and parts remain stable (Wanvik, 2014). All three dimensions underscore the pivotal changeability and constant emergence of assemblages, rather than their stability and permanence.

Employing the vocabulary of assemblage thinking allows us to better conceptualize the change and instability of carbonscapes. Instead of understanding the interweaving material, social and political structures or socio-technical regimes of "fossil capitalism" (Huber, 2013) as stable totalities, this vocabulary enables us to theorize the interlinkages between stability and change: Carbonscapes have, since the industrial revolution, been subject to powerful *territorialization* processes that have embedded fossil-based infrastructures, both materially and socially. Since the Second World War, the suburbanization of many cities in the Global North has put in place sprawling urban form, in a material sense. Yet, this has also bound conceptions of freedom and wealth together with high energy consumption, the private car, and the larger political–institutional and corporate structures of global oil markets. This *carbonscape assemblage* is strongly territorialized across a range of social and material processes and artefacts.

At the same time, the carbonscape assemblage is made inherently unstable through its relations of exteriority. It has no core, no essence and no fixed identity holding it together. The various elements through which it is composed—the political–institutional structures, the global oil markets, the material infrastructures and the socio-cultural discourses of freedom—are themselves integrated with other assemblages which subject them to specific pressures.

5.4   The Instabilities of Contested Energy Spaces

Contested energy spaces contain certain landscape features. There is a strong tradition within geography of seeing landscapes as more than material artefacts but rather as permeated by, or constructed through, social, political, cultural and economic relationships (Mitchell, 2002). As Zukin et al. (1992, p. 221) explained, landscapes are "built around dominant social institutions [...] and ordered by their power". But, even though the material realm is often associated with structure and inertia, the time–space of assemblages is imagined as inherently unstable and infused with movement and change (Marcus & Saka, 2006; Scott-Cato & Hillier, 2010). Assemblages and their component parts are recognized by their emergent capacities, properties that are contingent by their interaction with other component parts (DeLanda, 2016; DeLanda & Harman, 2017). Instead of seeing social entities as totalities (organic wholes bound together by internal relations), assemblage theory stresses how the interactions between seemingly separate elements produce unstable and contingent entities, revealing the empirical stability of carbonscapes as temporary, contingent achievements, always vulnerable to reconfigurations.

Employing DeLanda's notion of an assemblage "converter" (DeLanda, 2016; Deleuze & Guattari, 1987, pp. 324–325) might highlight the catalytic impact of well-placed component parts in either transforming assemblages or ensuring that relations and parts remain stable (Haarstad & Wanvik, 2016; Wanvik, 2014). These converters can theoretically ascend from everywhere within or outside of the assemblage. However, component parts of carbonscape assemblages can serve as particularly influential, transformative forces when they happen to be well-placed to create ripple effects. Hence, employing the vocabulary of assemblage theory allows us to better conceptualize the change and instability of carbonscapes. In the following, I will elaborate on how this instability plays out empirically by scrutinizing landscape instability, governance instability and power instability created within the contested energy space assemblage of the Canadian oil sands region.

Landscape Instability

The material consequences of resource extraction and the subsequent environmental degradation and the limited, non-renewable character of the energy resource are all defining characteristics of carbonscapes (Le Billon & Carter, 2012; Marsden, 2010; Nikiforuk, 2010). The hostile features of the Albertan carbonscape, with its vast tailing ponds, huge open wounds and crisscrossing seismic lines in the boreal forest landscape, contribute to its vulnerable and unstable characteristics.

The boreal forest is the dominant forest region in Canada, making up to 35% of Canada's landmass and 77% of the country's forested land. It is also one of the world's largest ecosystems, comprising 10% of the world's forest cover. The boreal forest plays a vital role in sustaining ecological cycles, including ecological services like wildlife diversity and carbon storage. It is an essential part of the environment, contributing to healthy air, water and soil, and is also a vital economic resource for the region (Sweeney, 2012, p. 13).

The boreal forest is a dynamic system, subject to highly transformative processes. Among them, wildfires are natural parts of its life cycle; many of the vegetation species are well adapted to large wildfires. Wildfires consume forest canopy and can spread from treetop to treetop, releasing huge quantities of sparks, smoke and other gases (Sweeney, 2012). Before major wildfire suppression programs, boreal forests historically burned on an average cycle ranging from 50 to 200 years as a result of lightning and human-caused wildfires, being part of indigenous controlled burning practices (Lewis & Ferguson, 1988; McCormack, 2017). However, due to reduced wildfire activity in the wake of industrial developments, forests of Alberta are ageing, which ultimately changes ecosystems and increases the risk of large and potentially costly catastrophic, unstable wildfires (McCormack, 2017; Sweeney, 2012, p. 13). In the third case study, I elaborate on how one such devastating wildfire event leads to substantial political consequences (see Chap. 8).

Governance Instability

Although the landscape of the oil sands region is subject to abrupt changes, the social and political features of the Wood Buffalo carbonscape are equally prone to instability and rupture. The Regional Municipality of Wood Buffalo (RMWB) is a specialized municipality located in north-eastern Alberta. Formed as a result of the amalgamation of the City of Fort McMurray and Improvement District No. 143 in 1995, it is the second largest municipality in Alberta by area. It is home to the vast Athabasca oil sand deposits, making the region one of the fastest-growing industrial areas in Canada. Rural populations of primarily indigenous peoples in the RMWB have expressed that they experience most of the negative impacts from industrial developments (i.e. environmental degradation, influx of foreign specialized labour and increased heavy transport activity), while the urban area of Fort McMurray benefits from most of the tax revenues (McDermott, 2016). This internal rural–urban politics and the inequalities that have been seen in the amalgamation and since, within the RMWB will be a key focus in the following analysis.

With their traditionally shared interest in smooth operations of extractive industries, government and industry claim to make strong efforts to include and integrate indigenous communities in the value chain to enable them to benefit from the positive impacts of industrial developments (Brownsey & Rayner, 2009). These efforts have materialized through extensive consultation processes, environmental impact assessments and impact benefit agreements (see first case study, Chap. 6 for details). Through these processes, government and industry are managing, maintaining and stabilizing the contested energy spaces of the oil sands. However, today most of these territorializing processes are delegated to industry, framed as corporate social responsibility (CSR) and stakeholder management (Fidler, 2010; Harvey & Bice, 2014; Lawrence & Macklem, 2000; Morgan, 2012; O'Faircheallaigh, 2013; Wanvik, 2016). Turning governance into CSR, the governance regime within the Wood Buffalo carbonscape assemblage is greatly dependent on some precarious externalities influencing corporate profits and margins, to maintain a certain level of infrastructure and services (Wanvik, 2016). This makes the governance of carbonscapes highly vulnerable to external shocks. The recent drop in global oil prices has put tens of thousands out of work and reduced government revenues substantially (Alberta Energy, 2017; Alberta Government, 2017). Converters in one part of the oil assemblage (oil price, and geopolitical conflicts) have led to major disruptions throughout the carbonscape on the global scale. The economic downturn hitting the price of oil in 2014 was an important background for the political negotiations that followed in the wake of the Fort McMurray wildfire.

Power Instability

For decades, the asymmetrical context of the Canadian North, with its economically disadvantaged rural Indigenous communities and explosive economic growth in some urban cores (fuelled by extractive industries), has been a concern for social scientists (Angell & Parkins, 2011). Recognizing the limitations of a "passive vic-tim" perspective on Indigenous peoples, contemporary scholars have advocated a new research perspective, one that is more responsive to the changing milieu of northern Indigenous peoples and that "recognises Indigenous peoples as conscious, pragmatic actors in cultural change and adaptation" (Angell & Parkins, 2011, p. 72). The call for a new approach to northern Indigenous research stems from the grow-ing political power among northern peoples (Hovelsrud & Krupnik, 2006). A better understanding of Indigenous people's mobilization in support of their own goals and aspirations is required, including information about how they proactively respond to industrial developments and other forms of externalities as agents in their own right (Alcantara, Lalonde, & Wilson, 2017), rather than solely focusing on their reactive resistance. Recent literature has also put focus on the effectiveness of Indigenous multi-community organization under the threat of oil development in Alberta (Longley, 2015), adding to the richness of scholarly contributions to under-standing Indigenous mobilization of resources.

Elsewhere, I argue that an analytical framework is needed to properly understand and explain recent developments among Métis communities in northern Alberta (Wanvik & Caine, 2017). Several Métis communities in this region have mobilized a variety of resources to increase their leverage and expand their rights in the midst of the oil sands frenzy. Such transformative competence is nothing new (Pelling, 2011), but it has not been fully elaborated through studies of Indigenous resource politics. Hence, rather than being subject to circumstance, I argue that Indigenous communities seize the moment through strategic and pragmatic engagement with an ever-changing environment (see Chap. 7 for details). The alliance forged among Wood Buffalo Métis communities was instrumental in mobilizing Indigenous and rural people during the Fort McMurray wildfire, and played a vital part in the politi-cal negotiations that followed (see last case study, Chap. 8 for details).

5.5   Conclusion

From this perspective, transformation is not dependent upon some future overthrow of the "system as a whole". Change always occurs in particular assemblages by way of reconfiguration, adaptation and conversion. Dramatic changes in one assemblage can destabilize other assemblages to which it is attached. Assemblages can have *emergent capacities* for change that are difficult to see because change is contingent upon interaction with other component parts. Our purpose here is not to argue for a complete adoption of DeLanda's assemblage thinking by energy geography (that would go against the very intention of thinking with assemblages). However, moving away from theoretically constructing near-total coherence around the elements of "petroculture" (Marriott & Minio-Paluello, 2012), "fossil capitalism" (Huber, 2013), "carbon democracy" (Mitchell, 2011) helps us theorize and visualize change processes and potentials. Assemblage thinking provides us with an effective social ontology and a vocabulary for this purpose.

I would argue that there is a need to reconceptualize the stabilities and instabilities of fossil fuel-based societies in ways that reveal new pathways for change and transformation. This should in no way disregard the significant permanence created by the embeddedness of energy in various aspects of society, which would obviously overlook important historical experience. However, if we are to understand how stabilities interact with volatility and instabilities—which are also important aspects of historical experience, then we need theoretical frameworks that enable us to identify and analyse them. In the next section, in the first of three case studies, I employ three different analytical lenses through which to examine these instabilities.

References

Alberta Energy. (2017). *Resource revenues collected.* Retrieved from http://www.energy.alberta.ca/About_Us/2564.asp

Alberta Government. (2017). *Unemployment rate.* Retrieved from http://economicdashboard.alberta.ca/Unemployment

Alcantara, C., Lalonde, D., & Wilson, G. N. (2017). Indigenous research and academic freedom: A view from political scientists. *International Indigenous Policy Journal, 8*(2), 3.

Allen, J. (2011). Powerful assemblages? *Area, 43*(2), 154–157. https://doi.org/10.1111/j.1475-4762.2011.01005.x

Anderson, B., Kearnes, M., McFarlane, C., & Swanton, D. (2012). On assemblages and geography. *Dialogues in Human Geography, 2,* 171–189. https://doi.org/10.1177/2043820612449261

Angell, A. C., & Parkins, J. R. (2011). Resource development and aboriginal culture in the Canadian north. *Polar Record, 47*(1), 67–79. https://doi.org/10.1017/S0032247410000124

Bakker, K., & Bridge, G. (2006). Material worlds? Resource geographies and the matter of nature'. *Progress in Human Geography, 30*(1), 5–27.

Bebbington, A., Hinojosa, L., Bebbington, D. H., Burneo, M. L., & Warnaars, X. (2008). Contention and ambiguity: Mining and the possibilities of development. *Development and Change, 39*(6), 887–914.

Bennett, J. (2005). The agency of assemblages and the North American blackout. *Public Culture, 17*(3), 445.

Boyer, D. (2011). Energopolitics and the anthropology of energy. *Anthropology News, 52*(5), 5–7.

Bridge, G. (2017). The map is not the territory: A sympathetic critique of energy research's spatial turn. *Energy Research & Social Science, 36*, 11.

Bridge, G., Bouzarovski, S., Bradshaw, M., & Eyre, N. (2013). Geographies of energy transition: Space, place and the low-carbon economy. *Energy Policy, 53*, 331–340.

Brownsey, K., & Rayner, J. (2009). Integrated land management in Alberta: From economic to environmental integration. *Policy and Society, 28*(2), 125–137.

Bulkeley, H., & Betsill, M. M. (2013). Revisiting the urban politics of climate change. *Environmental Politics, 22*(1), 136–154.

Bulkeley, H., Castán Broto, V., & Maassen, A. (2013). Governing low carbon transitions. In H. Bulkeley, V. Castán Broto, M. Hodson, & S. Marvin (Eds.), *Cities and low carbon transitions* (pp. 29–41). London, UK: Routledge.

Bulkeley, H., Castán Broto, V., & Maassen, A. (2014). Low-carbon transitions and the reconfiguration of urban infrastructure. *Urban Studies, 51*(7), 1471–1486.

Calvert, K. (2016). From 'energy geography' to 'energy geographies' Perspectives on a fertile academic borderland. *Progress in Human Geography, 40*(1), 105–125.

Coenen, L., Benneworth, P., & Truffler, B. (2012). Toward a spatial perspective on sustainability transitions. *Research Policy, 41*, 968–979.

DeLanda, M. (2006). *A new philosophy of society: Assemblage theory and social complexity.* London, UK: A&C Black.

DeLanda, M. (2016). *Assemblage theory.* Edinburgh, Scotland: Edinburgh University Press.

DeLanda, M., & Harman, G. (2017). *The rise of realism.* Cambridge, UK: Polity Press.

Deleuze, G., & Guattari, F. (1987). *A thousand plateaux* (B. Massumi, Trans.). Minneapolis, MN: University of Minnesota Press.

Deleuze, G., & Guattari, F. (1988). *A thousand plateaus: Capitalism and schizophrenia.* London, UK: Bloomsbury Publishing.

Deleuze, G., & Parnet, C. (2007 [1977]). Dialogues II (Revised ed). New York, NY: Columbia University Press.

Fidler, C. (2010). Increasing the sustainability of a resource development: Aboriginal engagement and negotiated agreements. *Environment, Development and Sustainability, 12*(2), 233–244. https://doi.org/10.1007/s10668-009-9191-6

Gibson-Graham, J. K. (2006). *A postcapitalist politics.* Minneapolis, MN: University of Minnesota Press.

Gidwani, V. K. (2008). *Capital, interrupted: Agrarian development and the politics of work in India.* Minneapolis, MN: University of Minnesota Press.

Haarstad, H. (2012). Extracting justice? Critical themes and challenges in Latin American natural resource governance. In *New political spaces in Latin American natural resource governance* (pp. 1–16). Berlin, Germany: Springer.

Haarstad, H. (2014a). Climate change, environmental governance and the scale problem. *Geography Compass, 8*(2), 87–97.

Haarstad, H. (2014b). Cross-scalar dynamics of the resource curse: Constraints on local participation in the Bolivian gas sector. *Journal of Development Studies, 50*(7), 977–990.

Haarstad, H., & Fløysand, A. (2007). Globalization and the power of rescaled narratives: A case of opposition to mining in Tambogrande, Peru. *Political Geography, 26*(3), 289–308.

Haarstad, H., & Wanvik, T. I. (2016). Carbonscapes and beyond conceptualizing the instability of oil landscapes. *Progress in Human Geography, 41*, 432. https://doi.org/10.1177/0309132516648007

Hansen, T., & Coenen, L. (2015). The geography of sustainability transitions: Review, synthesis and reflections on an emergent research field. *Environmental Innovation and Societal Transitions, 17*, 92–109.

Harvey, B., & Bice, S. (2014). Social impact assessment, social development programmes and social licence to operate: Tensions and contradictions in intent and practice in the extractive sector. *Impact Assessment and Project Appraisal, 32*(4), 327–335. https://doi.org/10.1080/14 615517.2014.950123

Harvey, D. (1989). From managerialism to entrepreneurialism: The transformation in urban governance in late capitalism. *Geografiska Annaler, 71B*(1), 3–17.

Hodson, M., & Marvin, S. (2010). Can cities shape socio-technical transitions and how would we know if they were? *Research Policy, 39*, 477–485.

Hommels, A. (2005). Studying obduracy in the city: Toward a productive fushion between technology studies and urban studies. *Science, Technology, & Human Values, 30*(3), 323–351.

Hovelsrud, G. K., & Krupnik, I. (2006). IPY 2007-08 and social/human sciences: An update. *Arctic, 59*(3), 341–348.

Huber, M. T. (2013). *Lifeblood: Oil, freedom, and the forces of capital*. Minneapolis, MN: University of Minnesota Press.

Humphreys, M., Sachs, J. D., & Stiglitz, J. E. (2007). Introduction: What is the problem with natural resource wealth? In M. Humphreys, J. D. Sachs, & J. E. Stiglitz (Eds.), *Escaping the resource curse* (pp. 1–20). New York, NY: Columbia University Press.

Kander, A., Malanima, P., & Warde, P. (2014). *Power to the people: Energy in Europe over the last five centuries*. Princeton, NJ: Princeton University Press.

Keeling, A., & Sandlos, J. (2015). *Mining and communities in Northern Canada: History, politics, and memory*. Calgary, Canada: University of Calgary Press.

Kemp, R., Schot, J., & Hoogma, R. (1998). Regime shifts to sustainability through processes of niche formation: The approach of strategic niche management. *Technology Analysis & Strategic Management, 10*(2), 175–198.

Kirshner, J., & Power, M. (2015). Mining and extractive urbanism: Postdevelopment in a Mozambican boomtown. *Geoforum, 61*, 67–78.

Latour, B. (2005). Reassembling the social-an introduction to actor-Network-theory. *Reassembling the social-an introduction to actor-network-theory* (p. 316). Foreword by Bruno Latour. Oxford, UK: Oxford University Press. ISBN-10: 0199256047. ISBN-13: 9780199256044, 1.

Lawrence, S., & Macklem, P. (2000). From consultation to reconciliation: Aboriginal rights and the Crown's duty to consult. *Canadian Bar Review, 29*, 252–279.

Le Billon, P., & Carter, A. (2012). Securing Alberta's tar sands: Resistance and criminalization on a new energy frontier. In *Natural resources and social conflict: Towards critical environmental security* (pp. 170–192). Basingstoke, UK: Palgrave Macmillan.

Lewis, H. T., & Ferguson, T. A. (1988). Yards, corridors, and mosaics: How to burn a boreal forest. *Human Ecology, 16*(1), 57–77.

Logan, O., & McNeish, J. (2012). Rethinking responsibility and governance in resource extraction. In J. McNeish & O. Logan (Eds.), *Flammable societies: Studies on the socio-economics of oil and gas* (pp. 1–44). London, UK: Pluto Press.

Longley, H. (2015). Indigenous battles for environmental protection and economic benefits during the commercialization of the Alberta oil sands. In A. Keeling & J. Sandlos (Eds.), *Mining and communities in Northern Canada: History, politics, and memory* (pp. 1967–1986). Calgary, Canada: University of Calgary Press.

Marcus, G. E., & Saka, E. (2006). Assemblage. *Theory, Culture & Society, 23*(2-3), 101–106.

Markard, J., Raven, R., & Truffler, B. (2012). Sustainability transitions: An emerging field of research and its prospects. *Research Policy, 41*, 955–967.

Marriott, J., & Minio-Paluello, M. (2012). *The oil road: Travels from the Caspian to the city*. New York, NY: Verso Books.

Marsden, W. (2010). *Stupid to the last drop: How Alberta is bringing environmental Armageddon to Canada (and doesn't seem to care)*. Toronto, Canada: Vintage Canada.

Martinez, N. (2007). *Bolivia's nationalization: Understanding the process and gauging the results*. Retrieved from Washington, DC.

McCormack, P. A. (2017). Walking the land: Aboriginal trails, cultural landscapes, and archaeological studies for impact assessment. *Archaeologies, 13*(1), 110–135.

McDermott, V. (2016, August 4). Rural hamlets asks province to examine amalgamation, News article. *Fort McMurray Today*. Retrieved from http://www.fortmcmurraytoday.com/2016/08/04/rural-hamlets-asks-province-to-examine-amalgamation

McFarlane, C., & Anderson, B. (2011). Thinking with assemblage. *Area, 43*(2), 162–164. https://doi.org/10.1111/j.1475-4762.2011.01012.x

McFarlane, C., & Rutherford, J. (2008). Political infrastructures: Governing and experiencing the fabric of the city. *International Journal of Urban and Regional Research, 32*(2), 363–374.

Mehlum, H., Moene, K., & Torvik, R. (2006). Institutions and the resource curse. *The Economic Journal, 116*(508), 1–20.

Mitchell, D. (2002). Cultural landscapes: The dialectical landscape-recent landscape research in human geography. *Progress in Human Geography, 26*(3), 381–390.

Mitchell, T. (2011). *Carbon democracy: Political power in the age of oil*. New York, NY: Verso Books.

Morgan, R. K. (2012). Environmental impact assessment: The state of the art. *Impact Assessment and Project Appraisal, 30*(1), 5–14. https://doi.org/10.1080/14615517.2012.661557

Müller, M. (2015). Assemblages and actor-networks: Rethinking socio-material power, politics and space. *Geography Compass, 9*(1), 27–41.

Nikiforuk, A. (2010). *Tar sands: Dirty oil and the future of a continent*. Vancouver, Canada: Greystone Books.

O'Faircheallaigh, C. (2013). Extractive industries and Indigenous peoples: A changing dynamic. *Journal of Rural Studies, 30*, 20–30.

Ogden, L., Heynen, N., Oslender, U., West, P., Kassam, K.-A., & Robbins, P. (2013). Global assemblages, resilience, and Earth Stewardship in the Anthropocene. *Frontiers in Ecology and the Environment, 11*(7), 341–347.

Ong, A. (2007). Neoliberalism as a mobile technology. *Transactions of the Institute of British Geographers, 32*(1), 3–8.

Pasqualetti, M. J., & Brown, M. A. (2014). Ancient discipline, modern concern: Geographers in the field of energy and society. *Energy Research & Social Science, 1*, 122–133.

Pelling, M. (2011). *Adaptation to climate change: From resilience to transformation*. London, UK: Routledge.

Perreault, T. (2006). From the Guerra del Agua to the Guerra del Gas: Resource governance, neoliberalism and popular protest in Bolivia. *Antipode, 38*(1), 150–172.

Pflieger, G., Kaufmann, V., Pattaroni, L., & Christophe, J. (2009). How does urban public transport change cities? Correlations between past and present transport and urban planning policies. *Urban Studies, 46*(7), 1421–1437.

Rip, A., & Kemp, R. (1998). Technological change. In S. Rayner & E. L. Malone (Eds.), *Human choice and climate change* (Vol. 2, pp. 327–399). Columbus, OH: Battelle Press.

Rutherford, J., & Coutard, O. (2014). Urban energy transitions: Places, processes and politics of socio-technical change. *Urban Studies, 51*(7), 1353–1377.

Sandlos, J., & Keeling, A. (2013). Living with zombie mines. *Seeing the Woods: A Blog*.

Sassen, S., & Ong, A. (2014). The carpenter and the bricoleur. In M. Acuto & C. Simon (Eds.), *Reassembling international theory: Assemblage thinking and international relations* (pp. 17–24). Basingstoke, UK: Palgrave Macmillian.

Scott-Cato, M., & Hillier, J. (2010). How could we study climate-related social innovation? Applying Deleuzean philosophy to Transition Towns. *Environmental Politics, 19*(6), 869–887.

Shove, E., Watson, M., & Spurling, N. (2015). Conceptualizing connections: Energy demand, infrastructures and social practices. *European Journal of Social Theory, 18*(3), 274–287.

Stevens, P., & Dietsche, E. (2008). Resource curse: An analysis of causes, experiences and possible ways forward. *Energy Policy, 36*(1), 56–65.

Sweeney, B. (2012). *Flat top complex wildfire review committee*. Retrieved from Alberta, Canada. http://wildfire.alberta.ca/wildfire-prevention-enforcement/wildfire-reviews/documents/FlatTopComplex-WildfireReviewCommittee-A-May18-2012.pdf

Tsing, A. L. (2005). *Friction: An ethnography of global connection*. Princeton, NJ: Princeton University Press.

Unruh, G. (2000). Understanding carbon lock-in. *Energy Policy, 28*(12), 817–830.

Urry, J. (2013). *Societies beyond oil: Oil dregs and social futures*. London, UK: Zed Books.

Urry, J. (2014). The problem of energy. *Theory, Culture & Society, 31*(5), 3–20.

Wanvik, T. I. (2014). Encountering a multidimensional assemblage: The case of Norwegian corporate social responsibility activities in Indonesia. *Norsk Geografisk Tidsskrift - Norwegian Journal of Geography, 68*(5), 282–290. https://doi.org/10.1080/00291951.2014.964761

Wanvik, T. I. (2016). Governance transformed into CSR – new governance innovations in the Canadian oil sands. *The Extractive Industries and Society, 3*, 517.

Wanvik, T. I., & Caine, K. (2017). Understanding indigenous strategic pragmatism: Métis engagement with extractive industry developments in the Canadian North. *The Extractive Industries and Society, 4*(3), 595–605. https://doi.org/10.1016/j.exis.2017.04.002

Watts, M. (2004). Resource curse? Governmentality, oil and power in the Niger Delta, Nigeria. *Geopolitics, 9*(1), 50–80.

Watts, M. (2013). Oil talk. *Development and Change, 44*(4), 1013–1026.

Zimmerer, K. S. (2011). New geographies of energy: Introduction to the special issue. *Annals of the Association of American Geographers, 101*(4), 705–711.

Zukin, S., Lash, S., & Friedman, J. (1992). *Postmodern urban landscapes: Mapping culture and power*. New York, NY.

Chapter 6
Case Study I: Governance of Contested Energy Spaces

Abstract In the contested space of energy production in Canada, tension and a series of disputes over land and rights have been prevalent between the state, industry and local Indigenous communities. Canadian governments have long exploited the natural resources of the land, while at the same time attempting to reconcile a difficult relationship with Indigenous communities living in proximity to the resources. This case study reveals how the government has conceded responsibility to industry to resolve the many governance challenges of Canada's energy spaces. Through substantial delegation of governance duties to industry, the Canadian Government has placed large parts of its regulatory toolbox in the hands of multinational companies, and hence turned social and environmental planning and programming into corporate stakeholder management.

Keywords Governance · Stakeholder management · Post-political · CSR

For decades, geographers have been studying encounters between global enterprises and local communities. In line with assemblage thinking (see previous chapter), the case study of the Norwegian oil company Statoil and its ventures in the contested energy spaces of Canada provides insights into the workings of social, material and historical realities, challenging the image of corporate social responsibility and its wider implications for societal governance (Wanvik, 2016).

The global energy market in general and the contested energy spaces more specifically are fertile ground for governance innovations; hence, they are excellent sites for studying emerging governance practices. The activities of the extractive industry have had substantial impact on the social, cultural and environmental realities in these spaces (Gamu, Le Billon, & Spiegel, 2015; LeClerc & Keeling, 2015; Veltmeyer & Bowles, 2014; Virah-Sawmy, 2015). Although there have been immense benefits for Canadian society, the burden borne by local ecosystems and Indigenous communities is substantial, which adds up to a prolonged historical conflict between the Crown and Indigenous peoples over rights and entitlements (Cairns, 2000; Veltmeyer & Bowles, 2014). The reciprocal arrangement between industry and government on the one hand and local communities on the other has been observed to be skewed, with insufficient contribution to local development and fulfilment of Indigenous rights and entitlements (Dembicki, 2012; Dow, 2012; Foster, 2008; Kelly et al., 2010).

Historically, the governance structure of Canada's contested energy spaces has been dominated by two groups of actors, namely state (governments and bureaucracies at federal, provincial and municipal levels) and industry (Hoberg & Phillips, 2011). Huge efforts have been invested by these two sectors in developing a previously uneconomic energy commodity (bitumen) into a highly profitable activity, resulting in a thriving industrial venture (Sherval, 2015). However, this has not come without cost; bitumen extraction has reinforced past grievances among local Indigenous communities (Wanvik, 2016), which have once again being deprived of their hard-earned access to traditional territories (Black, D'Arcy, & Weis, 2014; Huseman & Short, 2012; Jamasmie, 2014).

To cope with these potentially destabilizing conditions, the Crown has facilitated the emergence of flexible governance innovations, comprised of three tangible measures, namely consultation, environmental impact assessments (EIAs) and impact and benefit agreements (IBAs) (Griffin, 2012; Harvey & Bice, 2014; Lemos & Agrawal, 2006; Reich, 2008; Solomon, Katz, & Lovel, 2008; Wanvik, 2016). These measures are based on the following objectives. All concerned parties are to be consulted and should make valuable contributions to governance processes; they should be invited to participate in the assessments of planned interventions; and they are expected to reach agreements based on certain minimum levels of consensus, the so-called positive-sum games or win–win solutions (Jacobsson & Garsten, 2012). Only through governance structures based on pragmatic "what works" criteria—the discourse goes—can proper management of the contested energy spaces be exercised (Jones, 2008).

Accordingly, industry has been delegated extensive responsibilities for the use of governance instruments. By encouraging companies to comply with international CSR standards, *and* by facilitating "beyond compliance activities", the Alberta government has put private interests in the driver's seat of the governance framework of its contested energy spaces (Wanvik, 2016). To understand how these governance innovations emerged, and what their impact and consequences have been, we must examine recent developments in three different but related strands of thought in governance and corporate responsibility scholarship. The first relates to the conventional shift from government to governance, a development characterized by a move from hierarchical, representational government by institutions under majority rule, to more networked, egalitarian stakeholder relations based on alleged consensus (see, for example, Bingham, Nabatchi, & O'Leary, 2005; Braithwaite & Levi, 2003; Jessop, 1997; Jones, 1998; Rhodes, 1997, 2007). The second development relates to the first, but goes further by identifying a radically interpreted, particular post-political condition, namely the emergence of a managerial, elitist space emptied of politics where decisions are based on pragmatic "what works" criteria (see, for example, Agamben et al., 2009; Brown, 2005; Crouch, 2000; Mouffe, 1999, 2005; Swyngedouw, 2005, 2011; Zizek, 1999). The third feature is the incremental evolution of corporate responsibility towards stakeholder management. This change in corporate practices can be viewed as a response to changing governing preferences, together with an increased maturity in corporate responsibility implementation, primarily among multinational companies, where traditional philanthropic, standardized and image-based corporate social responsibility (CSR) has been replaced by an allegedly collaborative, performance-driven and integrated practice (Brammer, Jackson, & Matten, 2012; Dentchev, van Balen, & Haezendonck, 2015; Porter & Kramer, 2006; Scherer, Palazzo, & Matten, 2014; Solomon et al., 2008; Visser, 2013). This is in line with what scholars have identified as a more "inclusive business model" (Virah-Sawmy, 2015). By following Statoil's[1] journey into the vast prairies of Alberta, I show how companies have become an integral part of the new governance structure of Canada through their pragmatic quest for a social licence to operate (Wanvik, 2016). Multinational companies have encountered a highly politicized reality within contested energy spaces, and from a mix of formal consultations, corporate self-assessments and bilateral negotiations, we see the emergence of hybrid governance structures, and more specifically the emergence of *governance as corporate stakeholder management*, in which industry plays a vital role.

6.1 Governance Theories and Corporate Responses

In the governance of complex societies, it appears to be a truism that there are a multitude of concerned parties. Theories of complex systems and networks have recently formed the basis of applied governance approaches to the participation of concerned parties or stakeholders (Dentchev et al., 2015; Dicken et al., 2001; Harvey & Bice, 2014; Jacobsson & Garsten, 2012; Jones, 1998; Reich, 2008; Sunley, 2008). The parties in such forms of governance participate (or are allowed to participate) in these decision-making relational networks because of their "stakes" in the issues that these forms of governance are intended to address (Swyngedouw, 2005). In the following, I revisit three basic notions behind the emergence of what I term *governance as corporate stakeholder management*. To understand these changing features of governance, we must examine three different but interrelated developments in the governance and management literature, combined with subsequent changes among corporate responsibility practitioners.

The concept of "governance" is used in geography and many other subdisciplines of the social sciences. Common elements emphasized are co-operation to enhance legitimacy, the effectiveness of governing societies, new processes and public–private arrangements (Kooiman, 2003). Traditionally, governing is what governments

---

[1] Statoil Canada Ltd. (Statoil) develops and operates the Kai Kos Dehseh (KKD) leases, which contain more than two billion barrels of estimated recoverable resources. Statoil employs more than 800 people, with its headquarters in Calgary, Alberta. Established by the Norwegian government in 1972, Statoil has grown to become one of Europe's leading oil and gas companies. The company operates 60% of all Norwegian oil and gas production (in Vaaland & Heide, 2008), and is Norway's largest single company with a net operating income of NOK 110 billion in 2014 (in Statoil, 2014).

do—they control the allocation of resources between social actors and provide a set of rules and operate a set of institutions to do so. Thus, governing involves the establishment of a basic set of relationships between governments and their citizens, which differ from highly structured and state-controlled hierarchical arrangements to those egalitarian or "plurilateral" society-driven ones that are monitored only loosely and informally, if at all (Howlett, Rayner, & Tollefson, 2009). In its broadest sense, "governance" is a term used to describe the mode of increased government co-ordination exercised by public *and* private actors in their efforts to solve problems of collective action inherent in government and governing (De Bruijn & Ernst, 1995; Klijn & Koppenjan, 2000; Kooiman, 2000; Rhodes, 1996). The driving force behind this development is said to be the increased recognition of societal complexity, and a growing awareness that governments are not the only crucial actor to address major societal challenges (Kooiman, 2000).

Governance of the contested energy spaces of Alberta can be said to be a process by which an ever-wider range of actors is drawn into governing processes thought to be characterized not by rules, regulations and the exercise of hierarchical authority, but by informal networks claimed to be egalitarian that focus upon partnerships and networks and the blurring of the boundaries between public and private sectors.

Although governance has gained considerable attention and endorsement, an influential group of scholars has strongly criticized its asserted crippling effects on democracy and participation (Mouffe, 2005; Rhodes, 2007; Swyngedouw, 2005, 2010). As Lemke (2007) points out, for all the positive aspects associated with the shift towards governance, there are also questions about its ability to improve democratic processes, not least about how it can potentially marginalize conflicts between groups or underplay contradictions between political objectives and actions—a condition referred to as "post-political" (Swyngedouw, 2010).

The post-political condition is held to be one where contestation and conflict are supplanted by consensus-based politics (Butler, Laclau, & Žižek, 2000). Central to this view is Mouffe's distinction between the "political" as the space of power, conflict and antagonism within human societies, and "politics", described as "the set of practices and institutions through which an order is created, organizing human coexistence in the context of conflictuality provided by the political" (Mouffe, 2005, p. 9).

Post-political analysis offers potentially useful insights into the framing of recent changes to governance systems, especially what is understood to be within the remit of governance, and who engages with a system and under what terms (Allmendinger & Haughton, 2012). According to Oosterlynck and Swyngedouw (2010), the new forms of governance that have arisen over the past decades have formed with the consensus—despite often conflicting agendas and lifestyles—that managerial–technological apparatuses should permit the negotiation of conflicts in such a way as to arrive at mutually beneficial policy formulations. This amounts to colonization of the political by managerial–technological governance that has erased the gap between the political and policies, resulting in depoliticization (Oosterlynck & Swyngedouw, 2010).

These changing spaces of governance correspond with developments within the field of corporate responsibility. The traditional concept of CSR is that corporations are a vital part of society, and that they have both the power and the responsibility to conduct their affairs in ways that satisfy not only shareholders but also other constituencies such as employees, customers, the environment and the community at large (Eijsbouts, 2011). Since the 1950s, CSR has increasingly become a buzzword in corporate–community relations. CSR has especially taken hold within extractive industries, first as a tool used by NGOs to police multinational mining or energy companies operating in the Global South (Dupuy, 2014; Harvey, 2014; Virah-Sawmy, 2015). Later, business gained control over its own CSR activities, leading to a proliferation of business-led CSR initiatives, concerning both international CSR standards and CSR reporting (Frynas, 2005; Harvey, 2014).

In the corporate responsibility literature, the principle of voluntarism is predominant and requires responsible business activities to be value based (Bowen & Johnson, 1953), discretionary and to extend beyond legal requirements (Carroll & Shabana, 2010; Dentchev et al., 2015; Eijsbouts, 2011; Lee, 2008). Among the many critics of voluntarism, Rajak (2011) states that CSR has evolved from a movement among campaigners to compel companies to "clean up their act" to a discourse of unity and partnership led by corporations themselves. Describing the historical development from activist-oriented naming and shaming of multinational companies to a more industry-led, self-inflicted social consciousness, Rajak claims that the moral economies of responsibility, generosity and community—and the social bonds of affection and coercion that these create—have become not the weapons of the weak, but the weapons of the powerful (ibid.).

A transformative concept within this critical development has been stakeholder management, a term first coined by Freeman in 1984 (2010). Maintaining "a licence to operate" is perceived to be a constant challenge (Harvey, 2014; Jenkins & Yakovleva, 2006; Virah-Sawmy, 2015), and for the extractive industries, corporate responsibility is about balancing the diverse demands of a wide array of stakeholders, with the ever-present need to make a profit (Jenkins, 2004). The stakeholder model was developed mainly by management scholars who were frustrated by the lack of practicality of the previous theoretical models. A stakeholder refers to any individual or group that maintains a stake in an organization in the way that a shareholder possesses shares. Furthermore, a stakeholder here is defined as any group or individual that "can affect or is affected by the achievement of an organization's objectives" (Freeman, 2010, p. 46). Within the stakeholder framework, the difference between the social and economic goals of a corporation is no longer relevant because the central issue is the survival and success of the corporation (Harvey, 2014; Lee, 2008; Virah-Sawmy, 2015). Survival of a corporation as such is affected not only by shareholders but also by various other stakeholders such as employees, governments and customers (Donaldson & Preston, 1995). Jones (1995) correctly predicted that the stakeholder model had great potential to become the central paradigm in the field of corporate responsibility.

In summary, I argue that the foundation upon which the governance innovations in contested energy spaces in Alberta have been constructed is based on three distinct but interwoven developments: (1) the theoretical and empirical evolution of governance as a multi-stakeholder approach that is more than government; (2) the subsequent critical identification of what has been termed a pragmatic, consensus-seeking, post-political condition and (3) the parallel emergence of corporate responsibility as stakeholder management. By adopting these theoretical lenses of analysis in the following case study, I provide a framework to explain how governance innovation is taking place in contested energy spaces.

6.2 Governance Innovations to Solve Tensions

To create a more stable relationship between the component parts of energy spaces, the government has developed flexible governance innovations, with three recognizable features, namely consultation, EIAs and IBAs. In the following, I elaborate on the genealogy of these different governance instruments, and how they can be said to constitute governance innovation reformulated as corporate stakeholder management. Each feature has its own academic and juridical literature, but the scope of this book does not allow me to engage with all of them. Hence, I offer only a brief, schematic outline of their characteristics, before elaborating on their empirical manifestations.

The Duty to Consult

Alberta energy spaces geographically overlap Indigenous traditional lands and treaty areas. The duty to consult is triggered by an apparent violation of an existing Indigenous or treaty right recognized and affirmed by the Constitution Act (1982), or in cases where Indigenous communities assert rights that have yet to be formally recognized by a court of law or treaty (Jacobsson & Garsten, 2012). This common law duty stems from the Crown's fiduciary obligation towards Indigenous peoples and Section 35 of the Constitution Act (1982), which are interrelated (Delgamuukw, 1997; Eyford, 2015; Gogal, Reigert, & Jamieson, 2005; Lawrence & Macklem, 2000). A number of court cases have debated and elaborated the infringements of Indigenous rights in relation to natural resource extraction, each contributing to the increasingly complex and multifaceted regulatory environment of the Canadian energy spaces (Gogal et al., 2005).

In most cases, this duty to consult is delegated to industry proponents. Crown policy often requires a private company to consult with adversely affected First Nations or other Indigenous communities (Delgamuukw, 1997; Haida, 2004; Sparrow, 1990; Tsilhqot'in, 2014). This delegation is pragmatically justified because the proponent has better insight into project details and is also best positioned to compensate for infringements (Alberta Government, 2014; Fidler, 2010; Gogal et al., 2005; Lawrence & Macklem, 2000). This practice of delegation is confirmed and backed by the Supreme Court of Canada (Gogal et al., 2005) and is also supported by local stakeholders: "Most communities would rather negotiate with companies than with the government or municipality" (interview with consultant, 2015). Although this is partly because of the historically bad relationship between Indigenous communities and the Crown, it is also because that the municipality has no resources to spare.

However, the duty to consult does not apply to Métis communities of this study because of the undecided consultative status of Métis communities in Alberta.[2] Alberta recognizes a duty to consult with some Métis communities when Crown land management and resource development decisions may adversely impact their traditional uses. "Currently, the province does not have a Métis consultation policy but has put in place an internal process to guide consultation with Métis communities on a case-by-case basis where there is a credible assertion of Métis rights" (email correspondence with community associated lawyer, 2014).

However, in the case of Métis communities and industry, the fact that consultation is delegated to industry is important: "Our stakeholder focus is on communities within 30 km of the facility" (Statoil Canada, 2007). Industry is less concerned about formalities regarding Indigenous status, and more worried about pragmatics: who are the stakeholders that can influence, or be influenced by, our performance? "We firmly believe that community consultation is a starting point for building the long-term sustainable relationships we need for successful oil sands development" (Statoil Canada, 2013).

EIA

The second governance feature in the Alberta energy spaces is EIAs. In practice, an EIA has a much broader scope than the duty to consult, but it contains some similar characteristics related to governance. As a key component of environmental management over the past 40 years, EIAs have coincided with the increasing recognition of the nature, scale and implications of environmental change brought about by human actions (Haida, 2004). As for the duty to consult, EIAs are delegated to companies (Morgan, 2012), but are carried out in close collaboration with government agencies.

Environmental disruptions are evident in the contested energy spaces, and local stakeholders report the material consequences of the industrial operations (Chipewyan Prairie Dené First Nation, 2007; Conklin Métis Local #193, 2012).[3] The material consequences are significant for Indigenous communities, which are

---

[2] Although, in its June 2007 Métis Harvesting Policy, Alberta conceded that Conklin is a rights-bearing community with harvesting rights (lawyer, legal analysis, appendix to traditional land use studies, Conklin Métis Local #193). Additionally, there are currently negotiations around an Albertan Métis consultation policy underway.

[3] These sources are traditional land use studies kindly provided by Chipewyan Prairie Dené First Nation and Conklin Métis Local #193. The sources contain data collected from elders in the two communities, based on interviews, field trips and storytelling.

surrounded by logging, exploration, development and production activities by both forestry and hydrocarbon industries: "It's no use, it will never get better or get back to the way it was before. The fish in the river are gone, the game has been driven out of these areas, and the few catches we get are sometimes rotten inside" (interview with elder, Fort McKay Métis Community, 2015).

These changes have affected subsistence practices and greatly impacted the freedom of Indigenous community members to move about the land for traditional land use purposes. The community is no longer free to hunt, trap, fish or gather berries and plants as it was previously (Chipewyan Prairie Dené First Nation, 2007; Conklin Métis Local #193, 2012; Connacher Inc., 2010). This mixed indigenous economy is in line with similar findings by other scholars: "For Indigenous communities, the mixed economy is dynamic and intrinsically bound to the environment, making the long-term impacts of industrial development especially critical" (LeClerc & Keeling, 2015, p. 17). The Statoil EIA for its Conklin project states:

"To ensure openness and transparency in the community, the company has undertaken a regional EIA that fully discloses the commercial development in the approximately 12 townships[4] of bitumen leases held by the company. This application and EIA discloses the development over the life of the project. The regional EIA regulatory approach was developed through consultation with provincial regulatory agencies" (Statoil Canada, 2007).

A common theme in most scholarly discussions of EIAs is a critique of the rationalist model of governance, pointing to the need to explore and develop models that embrace new thinking about planning and decision-making processes in their wider social, cultural, political and economic contexts (Alberta Government, 2013). This has encouraged the promotion of deliberative and collaborative approaches to planning and decision-making processes, including EIAs themselves, such as bringing stakeholders and communities into the processes, emphasizing the importance of communication as a means of negotiating consensus solutions that capture the values of those participants, and moving the professional technocrats from a controlling role to a facilitating role in the decision-making process (Bartlett & Kurian, 1999; Richardson, 2005; Wilkins, 2003). This view is supported by Statoil: "Several of the EIA programs, such as the wildlife monitoring for caribou, moose and wolf, were tailored to actively engage the local stakeholders and address their specific issues" (Statoil Canada, 2007). EIAs are commonly criticized for being biased in favour of proponents, and lacking peer-reviewed data analysis (Aguilar-Støen & Hirsch, 2015; Davidson & Mackendrick, 2004; Fidler, 2010; O'Faircheallaigh, 2007; Wilkins, 2003). Although Statoil's caribou monitoring project has been one of a rare collection of peer-reviewed EIA programs, it has also been the subject of substantial criticism for being too limited and unscientific (Boutin et al., 2012).

However, an EIA in itself is not constructed to mitigate or mediate environmental or social disturbances of planned industrial operations. For that purpose, there has been a proliferation of EIA follow-up initiatives, such as environmental agreements and other negotiated Impact Benefits Agreements (IBA), that are intended to reduce

---

[4]A standard geographical unit.

the widespread difficulty of ensuring effective follow-up of EIAs regarding both anticipated environmental impacts and their actual appearance (Elling, 2009; Wilkins, 2003), as well as to ensure monitoring to prepare for the unexpected.

IBA

IBA are privately negotiated agreements, typically between extractive industries and local communities, whereby government is relegated to an external observational role. IBAs are commonly viewed as agreements that establish formal relationships between signatories, mitigate negative development impacts and enhance positive development outcomes for Indigenous communities (O'Faircheallaigh, 2007). The agreements primarily focus on employment and economic benefits, while more recent IBA constructions acknowledge the need for greater flexibility and diversity of community involvement in industrial decisions and the need for social and cultural programs, dispute resolution mechanisms, revenue-sharing provisions and environmental restrictions (Caine & Krogman, 2010; Diges, 2008; Dreyer & Myers, 2005; Gibson, 2008; Sosa & Keenan, 2001).

IBAs are signed between extractive industries and Indigenous communities in Canada in general, and more specifically in Alberta, to establish formal relationships between them, to reduce the predicted impact of an industrial operation and to secure economic benefits for affected communities (Galbraith, Bradshaw, & Rutherford, 2007; Gibson, 2008; Sosa & Keenan, 2001). IBAs do not fall under the purview of the state and thus fall within a historically uncontested, grey area of legality, often referred to by lawyers as quasi-legal[5] (Sosa & Keenan, 2001; Wanvik, 2016).

To industry, these IBAs represent an opportunity to overcome a complicated situation resulting from the difficult relationship between the state and its Indigenous people:

"Government and the regulatory government for our industry; it's horrible, and it's got a lot of history to it. You have to appreciate that we [industry] are in the middle of this relationship, a nation-to-nation relationship. When they [Indigenous people] are not being recognized, all of that comes into the mix when industry tries to operate. And we have our own interests in doing things right" (interview with CSR manager, Statoil Canada, 2015).

This attitude resonates well with other parts of the industry, which claim that there may well be some very good business reasons for the extractive companies operating in frontier regions to want to pay attention and contribute to social development in their backyard. Companies should tune their operating models to help alleviate poverty, generate self-sustaining economic conditions that will drive the company's costs down over time and avoid community's unrest and criminal behaviour (Harvey, 2014, p. 8).

---

[5]Although they may become legally binding if the parties involved agree to this.

Conklin has negotiated a number of long-term agreements with industry: "These agreements have provided communities with direct funding support for physical, social, and human infrastructure, as well as contracting opportunities for company businesses and a process to address environmental issues involving future developments" (email correspondence with community lawyer, 2014). Clearly, IBAs avert the issue with regard to the consultative status of the Métis, making IBAs arguably the most useful of the three governance features for the Métis communities of Conklin and elsewhere. Since 2009, Conklin has taken an aggressive approach to asserting its Indigenous rights. Thus, the agreements signed by the community are comparable in their terms to those signed by local First Nations people. The agreements do not deal with compensation for the infringement of Indigenous rights, because that is a matter for which the Crown is completely responsible: "Rather, the money and business opportunities received from industry are intended to help the community cope with and respond to the massive change that oil sands activity is imposing on it" (email correspondence with community associated lawyer, 2014).

This approach is echoed in the statements from the company:

"We use a lot of resources in social investment, and I feel we have a good understanding of the situation that way. We operate in their back yard, so to speak, and want to be a good neighbour. We try to interact and compensate those who are affected by our operations in an adequate manner. There were a lot of social problems in Conklin, and many were thrilled when we arrived" (interview with CSR manager, Statoil Canada, 2014).

On signing an IBA, an Indigenous group accepts restrictions to the exercise of their Indigenous rights. They provide industry with access to their lands and give their support to the resource development project. In return, they accept a "package of measures" that include economic benefits and the minimization of negative impacts on the environment and people. Additionally, most IBAs contain provisions to ensure consent and co-operation from the Indigenous community, and confidentiality and non-compliance clauses (Caine & Krogman, 2010; Diges, 2008; Gogal et al., 2005; Keeping, 1999). Prno (2007) argues that Indigenous peoples find these agreements appealing because they lend legitimacy to Indigenous claims to land and rights (Caine & Krogman, 2010; Wanvik & Caine, 2017). Most communities recognize that the regulatory process is biased in favour of development, and communities seek economic and contracting benefits because "the choices they are faced with are either having development proceed and receiving some benefit from it, or having development proceed and receiving no benefit at all from it" (email correspondence with community associated lawyer, 2014).

Because of the grey area of legality concerning IBAs, there is some ambiguity regarding the claimed confidentiality surrounding these agreements. Some claim that "these agreements are generally kept confidential at the request of industry, since companies view them as business contracts, which under our legislation are entitled to confidentiality" (email correspondence with community associated lawyer, 2014), while industry claims that "Statoil will continue to honour our agreements with communities, and out of respect, that would include their confidentiality" (email correspondence with CSR manager, Statoil Canada, 2015). According to the

Government of Alberta's Aboriginal Consultation Office (ACO), social agreements (IBAs) have nothing to do with their duty to consult. They are confidential in nature, and there is nothing that compels companies or communities to divulge this information (email correspondence with head of ACO, 2015). However, Statoil noted: "Our agreements cover areas such as social investment, consultation, economic and workforce development. Our commitments are fairly generic and describe how we wish to work with our communities" (email correspondence with CSR manager, Statoil Canada, 2015). "We really want to link it to business risk. It is much more than being perceived as a good citizen of the world, there is a business rationale behind it" (interview with CSR advisor, Statoil Norway, 2015). "If we have healthier local communities benefiting from our programs, they are more loyal" (interview with CSR manager, Statoil Canada, 2015).

In summary, the rationale for Indigenous groups to enter into these agreements includes overcoming marginalization, strengthening regional economic and political sovereignty and increasing control of resources to ensure regional benefit flows returning to communities affected by development (Wanvik, 2016; Wanvik & Caine, 2017). Resource development proponents have an incentive to enter into IBAs with Indigenous communities to obtain consent from stakeholders to access the land for resource development, obtain labour locally and create a co-operative working relationship (Caine & Krogman, 2010; Wanvik, 2016).

6.3 Conclusion: The Emergence of Governance as Corporate Stakeholder Management

The extractive industry activities have had a great impact on the social, cultural and environmental realities in contested energy spaces of Alberta in general and in Conklin in particular. Environmentally, as well as socially and culturally, the burden shouldered by local ecosystems and Indigenous communities is substantial, and has added to a prolonged, historical conflict between the Crown and Indigenous peoples over rights and entitlements. This complex relationship has led to substantial challenges for all stakeholders in contested energy spaces.

In response to these challenges, the federal duty to consult along with provincial EIAs and locally negotiated IBAs have all been delegated to industry, representing component parts on different levels of a nested governance structure, where corporate responses in the form of corporate stakeholder management are positioned as an important centrepiece. This delegation has been legitimized on pragmatic grounds, underscoring industry's better positioning to consult the stakeholders, assess its own impact and negotiate compensation and benefit agreements. I have identified an interrelated, nested and multiscalar governance structure emerging from these four distinct governance features (consultations, EIAs, IBAs and stakeholder management) that can be viewed as a joint mobilizing effort by government, extractive industry proponents and Indigenous communities to realize a workable, win–win regulatory environment in the contested energy space (Fig. 6.1).

Fig. 6.1 Governance
transformed into CSR
(author's own graphic)

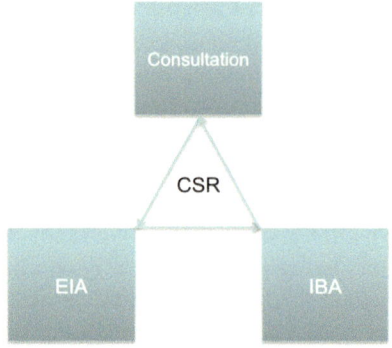

These are all recognizable features in the governance regime of contested energy spaces, where the emphasis is on a smooth transition from a highly political space—understood as Mouffe's space of power, conflict and antagonism—to governance, or rather processes, instruments and narratives such as "multi-stakeholderism", "community" and "partnership" (For elaborations on this issue, see Eyford, 2015; Gogal et al., 2005; Kennett, 1999). This structure is designed to govern energy spaces, where agreements are claimed to be in the mutual interest, where communities and corporations are rhetorically inseparable and where the survival of a company and the industry becomes the common objective for all stakeholders: "Without the oil sands, the community loses everything!" (Fort McKay Métis, 12.10.2015, social media update). Hence, notions of this governance practice as a positive-sum game are reinforced; the confidential nature of the IBAs ensures that this claimed mutual interest is upheld, turning local communities into silent, complacent stakeholders.

Consequently, the Alberta government has put private interests in the driver's seat in the governance of the energy space by encouraging companies to comply with international corporate responsibility standards, *and* by facilitating "beyond compliance activities" through the combination of delegating consultation with Indigenous communities to the companies, ensuring environmental impact (self-)assessments conducted by the companies, and letting the corporations negotiate IBAs bilaterally with the concerned communities. This incremental change in corporate practices can be viewed as a response to changing governing preferences, where negotiations, consensus and positive-sum games akin to a post-political condition are preferred to political competition over resources. This is based on a corresponding increased maturity in CSR implementation among primarily multinational companies. Here, traditional philanthropic, standardized and image-based corporate responsibility has been replaced by allegedly collaborative, performance-driven and integrated practice.

However, empirical evidence from this case study shows that risk management remains the central driver of stakeholder management. It remains to be determined whether industry has the resources and competencies to carry this acquired responsibility for local development actively and over time, and what happens when investments dry up and industry leaves. We can predict some potential shortcomings in the governance structure, particularly from its consent-producing IBAs.

With lucrative, confidential business agreements waiting at the end of a resource-demanding and tiring governance process, the possibility of bias in favour of industry development is high among the Indigenous communities in both consultations and EIAs (Wanvik & Caine, 2017). The role of government as regulator in this governance triangle is severely challenged by these bilaterally negotiated, confidential IBAs between industry and community. A first step towards a more transparent process should be to establish tripartite forums for these negotiations, where the local communities, the municipality responsible for local service delivery and the company sit down to agree on social investment needs and joint social programming.

However, the vital importance of stakeholder leverage in such negotiations also results in increased focus on documenting traditional land use among local Indigenous communities. This documentation is considered in conjunction with claims of cumulative environmental impact by existing and planned industrial developments: "Monitoring of prospect licensing by the government and mapping of historic and present traditional land use practices are important parts of our activities today" (interview with Métis consultant, 2015). Together with competence building related to negotiations, these activities are all part of the new reality of Indigenous communities. Hence, the communities themselves are calibrating their participatory role in the emerging governance processes in contested energy spaces to strengthen their negotiating power. In this way, they underscore the basic insight that there is no such thing as a post-political society.

References

Agamben, G., Badiou, A., Bensaïd, D., Brown, W., Nancy, J.-L., Rancière, J., et al. (2009). *Démocratie, dans quel état?* New York, NY: Cambridge University Press.

Aguilar-Støen, M., & Hirsch, C. (2015). Environmental impact assessments, local power and self-determination: The case of mining and hydropower development in Guatemala. *The Extractive Industries and Society, 2*(3), 472–479. https://doi.org/10.1016/j.exis.2015.03.001

Alberta Government. (2013). *Guide to preparing environmental impact assessment reports in Alberta.* Retrieved from Calgary, Alberta.

Alberta Government. (2014). *Government of Alberta proponent's guide to First Nations consultation procedures for land dispositions.* Retrieved from Calgary, Canada.

Allmendinger, P., & Haughton, G. (2012). Post-political spatial planning in England: A crisis of consensus? *Transactions of the Institute of British Geographers, 37*(1), 89–103.

Bartlett, R. V., & Kurian, P. A. (1999). The theory of environmental impact assessment: Implicit models of policy making. *Policy & Politics, 27*(4), 415–433.

Bingham, L. B., Nabatchi, T., & O'Leary, R. (2005). The new governance: Practices and processes for stakeholder and citizen participation in the work of government. *Public Administration Review, 65*(5), 547–558.

Black, T., D'Arcy, S., & Weis, T. (2014). *A line in the tar sands: Struggles for environmental justice.* Toronto, Canada: PM Press.

Boutin, S., Boyce, M. S., Hebblewhite, M., Hervieux, D., Knopff, K. H., Latham, M. C., … Serrouya, R. (2012). Why are caribou declining in the oil sands? *Frontiers in Ecology and the Environment, 10*(2), 65–67. https://doi.org/10.1890/12.WB.005

Bowen, H. R., & Johnson, F. E. (1953). *Social responsibility of the businessman*. New York, NY: Harper.

Braithwaite, V., & Levi, M. (2003). *Trust and governance*. New York, NY: Russell Sage Foundation.

Brammer, S., Jackson, G., & Matten, D. (2012). Corporate social responsibility and institutional theory: New perspectives on private governance. *Socio-Economic Review, 10*(1), 3–28. https://doi.org/10.1093/ser/mwr030

Brown, W. (2005). *Edgework. Critical essays on knowledge and politics*. Princeton, NJ: Princeton University Press.

Butler, J., Laclau, E., & Žižek, S. (2000). *Contingency, hegemony, universality: Contemporary dialogues on the left*. London, UK: Verso.

Caine, K. J., & Krogman, N. (2010). Powerful or just plain power-full? A power analysis of impact and benefit agreements in Canada's north. *Organization & Environment, 23*(1), 76–98. https://doi.org/10.1177/1086026609358969

Cairns, A. C. (2000). *Citizen plus - Aboriginal peoples and the Canadian state*. Vancouver, Canada: UBC Press.

Carroll, A. B., & Shabana, K. M. (2010). The business case for corporate social responsibility: A review of concepts, research and practice. *International Journal of Management Reviews, 12*(1), 85–105.

Chipewyan Prairie Dené First Nation. (2007). Kai'Kos'Dehseh Dené - The Red Willow River (Christina River) People. *A traditional land use study* (N. Press Ed., 1st ed.). Calgary, Alberta, Canada.

Conklin Métis Local #193. (2012). *Stories from a long time ago*. Retrieved from Conklin, Alberta, Canada.

Connacher Inc. (2010). *EIA: Appendix 7 - Traditional ecological knowledge and land use (TEK/TLU)*.

Crouch, C. (2000). *Coping with post-democracy* (Vol. 18). London, UK: Fabian Society.

Davidson, D. J., & Mackendrick, N. A. (2004). All dressed up with nowhere to go: The discourse of ecological modernization in Alberta, Canada. *Canadian Review of Sociology/Revue canadienne de sociologie, 41*(1), 47–65. https://doi.org/10.1111/j.1755-618X.2004.tb02169.x

De Bruijn, J. A., & Ernst, F. (1995). Policy networks and governance. In *Institutional design* (pp. 161–179). Dordrecht, Netherlands: Springer.

Delgamuukw v. British Columbia. (1997).

Dembicki, G. (2012). Oil sands carbon: When money and morals align. *The Tyee*.

Dentchev, N. A., van Balen, M., & Haezendonck, E. (2015). On voluntarism and the role of governments in CSR: Towards a contingency approach. *Business Ethics: A European Review, 24*, 378. https://doi.org/10.1111/beer.12088

Dicken, P., Kelly, P., Olds, K., & Yeung, H. W.-C. (2001). Chains and networks, territories and scales: Towards a relational framework for analysing the global economy. *Global Networks, 1*(2), 89–112.

Diges, C. (2008). *Sticks and bones: Is your IBA working? Amending and enforcing impact benefit agreements*. Toronto, Canada: McMillan Binch Mendelsohn LLP.

Donaldson, T., & Preston, L. E. (1995). The stakeholder theory of the corporation: Concepts, evidence, and implications. *Academy of Management Review, 20*(1), 65–91.

Dow, M. W. (2012). *Tarring the oil sands: The evolution and emergence of ENGO opposition in Alberta's oil sands and social movement theory*. Edmonton, Canada: University of Alberta.

Dreyer, D., & Myers, H. (2005). *Impact and benefits agreements: Do the Ross River Dena benefit from mineral projects?* Prince George, Canada: University of Northern British Columbia.

Dupuy, K. E. (2014). Community development requirements in mining laws. *The Extractive Industries and Society, 1*(2), 200–215.

Eijsbouts, J. (2011). *Corporate responsibility, beyond voluntarism* (Vol. Inaugural Lecture). Maastricht, Netherlands: Faculty of Law, Maastrich University.

Elling, B. (2009). Rationality and effectiveness: Does EIA/SEA treat them as synonyms? *Impact Assessment and Project Appraisal, 27*(2), 121–131.

Eyford, D. R. (2015). *A new direction - Advancing aboriginal and treaty rights.*

Fidler, C. (2010). Increasing the sustainability of a resource development: Aboriginal engagement and negotiated agreements. *Environment, Development and Sustainability, 12*(2), 233–244. https://doi.org/10.1007/s10668-009-9191-6

Foster, J. B. (2008). Peak oil and energy imperialism. *Monthly Review, 60*(3), 12.

Freeman, R. E. (2010). *Strategic management: A stakeholder approach.* New York, NY: Cambridge University Press.

Frynas, J. G. (2005). The false developmental promise of corporate social responsibility: Evidence from multinational oil companies. *International Affairs, 81,* 581–598.

Galbraith, L., Bradshaw, B., & Rutherford, M. B. (2007). Towards a new supraregulatory approach to environmental assessment in Northern Canada. *Impact Assessment and Project Appraisal, 25*(1), 27–41.

Gamu, J., Le Billon, P., & Spiegel, S. (2015). Extractive industries and poverty: A review of recent findings and linkage mechanisms. *The Extractive Industries and Society, 2*(1), 162–176. https://doi.org/10.1016/j.exis.2014.11.001

Gibson, V. V. (2008). *Negotiated spaces: Work, home and relationships in the Dene diamond economy.* Vancouver, Canada: University of British Columbia.

Gogal, S., Reigert, R., & Jamieson, J. (2005). Aboriginal impact and benefit agreements: Practical considerations. *Alberta Law Review, 43,* 129.

Griffin, L. (2012). Where is power in governance? Why geography matters in the theory of governance. *Political Studies Review, 10*(2), 208–220. https://doi.org/10.1111/j.1478-9302.2012.00260.x

Haida Nation v. British Columbia (Minister of Forests). (2004).

Harvey, B. (2014). Social development will not deliver social licence to operate for the extractive sector. *The Extractive Industries and Society, 1*(1), 7–11. https://doi.org/10.1016/j.exis.2013.11.001

Harvey, B., & Bice, S. (2014). Social impact assessment, social development programmes and social licence to operate: Tensions and contradictions in intent and practice in the extractive sector. *Impact Assessment and Project Appraisal, 32*(4), 327–335. https://doi.org/10.1080/14615517.2014.950123

Hoberg, G., & Phillips, J. (2011). Playing defence: Early responses to conflict expansion in the oil sands policy subsystem. *Canadian Journal of Political Science, 44*(3), 507–527.

Howlett, M., Rayner, J., & Tollefson, C. (2009). From government to governance in forest planning? Lessons from the case of the British Columbia Great Bear Rainforest initiative. *Forest Policy and Economics, 11*(5-6), 383–391. https://doi.org/10.1016/j.forpol.2009.01.003

Huseman, J., & Short, D. (2012). 'A slow industrial genocide': Tar sands and the indigenous peoples of northern Alberta. *The International Journal of Human Rights, 16*(1), 216–237.

Jacobsson, K., & Garsten, C. (2012). Post-political regulation: Soft power and post-political visions in global governance. *Critical Sociology, 39,* 421. https://doi.org/10.1177/0896920511413942

Jamasmie, C. (2014). Canada's oil sands brace for conflict-ridden 2014, News story. *Mining.com.* Retrieved from http://www.mining.com/canadas-oil-sands-brace-for-conflict-ridden-2014-95551/

Jenkins, H. (2004). Corporate social responsibility and the mining industry: Conflicts and constructs. *Corporate Social Responsibility and Environmental Management, 11*(1), 23–34.

Jenkins, H., & Yakovleva, N. (2006). Corporate social responsibility in the mining industry: Exploring trends in social and environmental disclosure. *Journal of Cleaner Production, 14*(3), 271–284.

Jessop, B. (1997). Capitalism and its future: Remarks on regulation, government and governance. *Review of International Political Economy, 4*(3), 561–581.

Jones, M. (1998). Restructuring the local state: Economic governance or social regulation? *Political Geography, 17*(8), 959–988. https://doi.org/10.1016/S0962-6298(97)00090-5

Jones, O. (2008). Stepping from the wreckage: Geography, pragmatism and anti-representational theory. *Geoforum, 39*(4), 1600–1612. https://doi.org/10.1016/j.geoforum.2007.10.003

Jones, T. M. (1995). Instrumental stakeholder theory: A synthesis of ethics and economics. *Academy of Management Review, 20*(2), 404–437.

Keeping, J. (1999). The legal and constitutional basis for benefits agreements: A summary. *Northern Perspectives, 25*(4).

Kelly, E. N., Schindler, D. W., Hodson, P. V., Short, J. W., Radmanovich, R., & Nielsen, C. C. (2010). Oil sands development contributes elements toxic at low concentrations to the Athabasca River and its tributaries. *Proceedings of the National Academy of Sciences, 107*(37), 16178–16183.

Kennett, S. A. (1999). *A guide to impact and benefits agreements.* Calgary, Canada: Canadian Institute of Resources Law.

Klijn, E.-H., & Koppenjan, J. F. (2000). Interactive decision making and representative democracy: Institutional collisions and solutions. In *Governance in modern society* (pp. 109–134). Dordrecht, Netherlands: Springer.

Kooiman, J. (2000). Societal governance : Levels, models, and orders of social-political interaction. In *Debating governance.* Oxford, UK: Oxford University Press.

Kooiman, J. (2003). *Governing as governance.* London, UK: Sage.

Lawrence, S., & Macklem, P. (2000). From consultation to reconciliation: Aboriginal rights and the Crown's duty to consult. *Canadian Bar Review, 29,* 252–279.

LeClerc, E., & Keeling, A. (2015). From cutlines to traplines: Post-industrial land use at the Pine Point mine. *The Extractive Industries and Society, 2*(1), 7–18. https://doi.org/10.1016/j.exis.2014.09.001

Lee, M.-D. P. (2008). A review of the theories of corporate social responsibility: Its evolutionary path and the road ahead. *International Journal of Management Reviews, 10*(1), 53–73. https://doi.org/10.1111/j.1468-2370.2007.00226.x

Lemke, T. (2007). An indigestible meal? Foucault, governmentality and state theory. *Distinktion: Scandinavian Journal of Social Theory, 8*(2), 43–64.

Lemos, M. C., & Agrawal, A. (2006). Environmental governance. *Annual Review of Environment and Resources, 31*(1), 297–325. https://doi.org/10.1146/annurev.energy.31.042605.135621

Morgan, R. K. (2012). Environmental impact assessment: The state of the art. *Impact Assessment and Project Appraisal, 30*(1), 5–14. https://doi.org/10.1080/14615517.2012.661557

Mouffe, C. (1999). Deliberative democracy or agonistic pluralism? *Social Research, 66*(3), 745–758. https://doi.org/10.2307/40971349

Mouffe, C. (2005). *On the political.* New York, NY: Psychology Press.

O'Faircheallaigh, C. (2007). Environmental agreements, EIA follow-up and aboriginal participation in environmental management: The Canadian experience. *Environmental Impact Assessment Review, 27*(4), 319–342. https://doi.org/10.1016/j.eiar.2006.12.002

Oosterlynck, S., & Swyngedouw, E. (2010). Noise reduction: The postpolitical quandary of night flights at Brussels airport. *Environment and Planning A, 42*(7), 1577.

Porter, M. E., & Kramer, M. R. (2006). The link between competitive advantage and corporate social responsibility. *Harvard Business Review, 84,* 78.

Prno, J. (2007). *Assessing the effectiveness of impact and benefit agreements from the perspective of their Aboriginal signatories.* Ann Arbor, MI: ProQuest.

R. v. Sparrow. (1990).

Rajak, D. (2011). *In good company: An anatomy of corporate social responsibility.* Stanford, CA: Stanford University Press.

Reich, R. (2008). Supercapitalism. The transformation of business, democracy and everyday life. *Society and Business Review, 3*(3), 256–258. https://doi.org/10.1108/sbr.2008.3.3.256.1

Rhodes, R. (1997). *Understanding governance. Policy networks, reflexivity and accountability.* Buckingham, UK: Open University Press.

Rhodes, R. (2007). Understanding governance: Ten years on. *Organization Studies, 28*(8), 1243–1264.

Rhodes, R. A. W. (1996). The new governance: Governing without government1. *Political Studies, 44*(4), 652–667.

Richardson, T. (2005). Environmental assessment and planning theory: Four short stories about power, multiple rationality, and ethics. *Environmental Impact Assessment Review, 25*(4), 341–365.

Scherer, A. G., Palazzo, G., & Matten, D. (2014). The business firm as a political actor: A new theory of the firm for a globalized world. *Business & Society, 53*(2), 143–156. https://doi.org/10.1177/0007650313511778

Sherval, M. (2015). Canada's oil sands: The mark of a new 'oil age' or a potential threat to Arctic security? *The Extractive Industries and Society, 2*(2), 225–236. https://doi.org/10.1016/j.exis.2015.01.011

Solomon, F., Katz, E., & Lovel, R. (2008). Social dimensions of mining: Research, policy and practice challenges for the minerals industry in Australia. *Resources Policy, 33*(3), 142–149. https://doi.org/10.1016/j.resourpol.2008.01.005

Sosa, I., & Keenan, K. (2001). *Impact benefit agreements between aboriginal communities and mining companies: Their use in Canada*. Toronto, Canada: Canadian Environmental Law Association.

Statoil. (2014). *Financial statements and review [Press release]*. Retrieved from http://www.statoil.com/no/InvestorCentre/QuarterlyResults/2014/Downloads/Financial%20statements%20and%20review%204Q%202014.pdf

Statoil Canada. (2007). *Application for approval of the Kai Kos Dehseh Project*. Retrieved from Calgary, Canada.

Statoil Canada. (2013). *Statoil community report 2013*.

Sunley, P. (2008). Relational economic geography: A partial understanding or a new paradigm? *Economic Geography, 84*(1), 1–26. https://doi.org/10.1111/j.1944-8287.2008.tb00389.x

Swyngedouw, E. (2005). Governance innovation and the citizen: The Janus face of governance-beyond-the-state. *Urban Studies, 42*(11), 1991–2006. https://doi.org/10.1080/00420980500279869

Swyngedouw, E. (2010). Impossible sustainability and the post-political condition. In *Making strategies in spatial planning* (pp. 185–205). Dordrecht, Netherlands: Springer.

Swyngedouw, E. (2011). Interrogating post-democratization: Reclaiming egalitarian political spaces. *Political Geography, 30*(7), 370–380.

Tsilhqot'in Nation v. British Columbia. (2014).

Vaaland, T. I., & Heide, M. (2008). Managing corporate social responsibility: Lessons from the oil industry. *Corporate Communications: An International Journal, 13*(2), 212–225. https://doi.org/10.1108/13563280810869622

Veltmeyer, H., & Bowles, P. (2014). Extractivist resistance: The case of the Enbridge oil pipeline project in Northern British Columbia. *The Extractive Industries and Society, 1*(1), 59–68. https://doi.org/10.1016/j.exis.2014.02.002

Virah-Sawmy, M. (2015). Growing inclusive business models in the extractive industries: Demonstrating a smart concept to scale up positive social impacts. *The Extractive Industries and Society, 2*(4), 676–679.

Visser, W. (2013). *CSR 2.0: Transforming corporate sustainability and responsibility*. Heidelberg, Germany: Springer.

Wanvik, T. I. (2016). Governance transformed into CSR – new governance innovations in the Canadian oil sands. *The Extractive Industries and Society, 3*, 517.

Wanvik, T. I., & Caine, K. (2017). Understanding indigenous strategic pragmatism: Métis engagement with extractive industry developments in the Canadian North. *The Extractive Industries and Society, 4*(3), 595–605. https://doi.org/10.1016/j.exis.2017.04.002

Wilkins, H. (2003). The need for subjectivity in EIA: Discourse as a tool for sustainable development. *Environmental Impact Assessment Review, 23*(4), 401–414.

Zizek, S. (1999). *The ticklish subject*. London, UK: Verso.

Chapter 7
Case Study II: Skewed Power Relations

Abstract This chapter challenges conventional understandings of Indigenous agency and responses to industrial activities on traditional Indigenous territories. By analysing the current mobilization of resources among three indigenous Métis communities in the regional municipality of Wood Buffalo, Alberta, this chapter develops an empirically grounded framework for understanding indigenous strategic pragmatism and the output, outcomes and impact of indigenous engagement with extractive industry developments.

Keywords Indigenous · Canada · Governance · Pragmatism · Assemblage · Mobilizing · Network

Although an understanding of the empowerment of indigenous communities facing extractive industrial developments is emerging, most academic research still favours conventional conceptualizations of local indigenous communities as subject to circumstance and who are pushed even further to the fringes of their lands by external forces threatening to extinguish their traditional ways of life. However, this conventional understanding of industrial-indigenous relations does not explain recent developments in the indigenous Metis communities of northern Alberta (Wanvik & Caine, 2017). Several of these indigenous communities have mobilized a variety of resources to increase their leverage and expand their rights in the midst of the oil sands. Rather than being subject to circumstance, in this chapter I argue that indigenous communities often seize the moment through *strategic* and *pragmatic* engagement with their ever-changing environments.

Typically, traditional indigenous cultures in the Canadian North are characterized as a "unique set of beliefs and practices, which have successfully sustained aboriginal peoples physically, socially, and spiritually, since time immemorial"

© The Author(s) 2018, under exclusive licence to Springer International Publishing AG, part of Springer Nature 2019
T. I. Wanvik, *Contested Energy Spaces*, SpringerBriefs in Geography, https://doi.org/10.1007/978-3-030-02396-6_7

(Angell & Parkins, 2011).[1] External disturbances, often represented by colonial or post-colonial industrial extraction activities and impacts on traditional territories (Justus & Simonetta, 1979, 1982), have been associated with cultural discontinuity leading to high rates of depression, alcoholism, suicide and violence in many indigenous communities (Kirmayer, Brass, & Tait, 2000).

7.1 Conventional Understanding of Indigenous Agency

As we have seen in the previous chapter, governance of relationships between extractive industries and Indigenous people is characterized by the comprehensive delegation of power from state institutions to industry (Arena, Bozzolan, & Michelon, 2015; Caine & Krogman, 2010; Fidler, 2010; Harvey & Bice, 2014; Lawrence & Macklem, 2000; O'Faircheallaigh, 2007; Prno & Scott Slocombe, 2012; Wanvik, 2016). For decades, the asymmetrical context of the Canadian North, with its economically disadvantaged rural Indigenous communities and explosive economic growth in urban cores (fuelled by ever-expanding extractive industries), has been a concern for social scientists from all disciplines (for a comprehensive review of this literature, see Angell & Parkins, 2011). Scholarship on Indigenous responses to extraction industries in the Canadian North can be described as a continuum with two distinct phases: the *community impact* period (1970–1995), which was marked by emphases on social pathologies and social disruption, the politics of assimilation, the sociology of disturbance and the anthropology of acculturation (see, for example, Erikson, 1976; Justus & Simonetta, 1979; Waldrum, 1988); and the *community continuity* period (1996 to present), which underscored the growing political empowerment of indigenous communities through cultural resistance (Barker, 2015), political inclusion in participatory governance processes (Fidler, 2010; Gibson & Klinck, 2005; Harvey & Bice, 2014; Lawrence & Macklem, 2000; O'Faircheallaigh, 1999, 2007, 2010a, 2010b) and attention to traditional knowledge in aboriginal communities (Usher, 2000).

I claim this continuous character of these approaches for several reasons. First, the asymmetrical character of the relationship between extractive industry proponents (including the Canadian State apparatus) and indigenous communities surrounding the industrial activities has been seen as skewed and oppressive (Hoberg & Phillips, 2011; Huseman & Short, 2012; Le Billon & Carter, 2012). Second, the cumulative effects of oil sands developments in Alberta from the 1960s to the 1980s

---

[1] In this chapter I primarily use the term "indigenous", but I also use "aboriginal" when it explicitly refers to documents or when groups use it to describe themselves. "Aboriginal Peoples" is the collective term for Métis, First Nations and Inuit that has been widely adopted by the Canadian government and many indigenous groups, and it was sanctified by law in Section 35(2) of the *Constitution Act, 1982*. The Canadian government now acknowledges the term "Indigenous Peoples" and recognizes their international legal rights under the United Nations Declaration of the Rights of Indigenous Peoples.

were extremely disruptive, hence there is good reason to consider community impacts (Justus & Simonetta, 1982). Industry continues to have negative environmental and social impacts for the local indigenous communities, who are losing trapping areas, seeing declines in traditional activities, experiencing the deterioration of their community's social fabric and losing the ability to control their community life (Conklin Métis Local #193, 2012; Fort McKay Sustainability Centre (TLU), 2016; McMurray Metis (TLU), 2015). Third, central to recent scholarship on indigenous responses to extractive developments is the concept of resilience, which explains how externally imposed impacts are experienced and buffered (Gibson & Klinck, 2005). Magis (2010) suggests that resilience pertains to the ability of a system to sustain itself through change via adaptation and occasional transformation, underscoring Healy's definition of indigenous community resilience as the capacity of a distinct community or cultural system to absorb disturbance and reorganize while undergoing change so as to retain key elements of structure and identity that preserve its distinctness (Fleming & Ledogar, 2008; Healey, 2006).

Recognizing the limitations of a "passive victim" research perspective, contemporary scholars have advocated a new research perspective, one that is more responsive to the changing milieu of northern Indigenous peoples and that "recognises Indigenous peoples as conscious, pragmatic actors in cultural change and adaptation" (Angell & Parkins, 2011, p. 72). The call for a new approach to northern Indigenous research stems from the growing political power among northern peoples, their increasing education levels and political astuteness (Hovelsrud & Krupnik, 2006), and the subsequent resurgence of Indigenous communities (Barker, 2015; Coates, 2015; Wotherspoon & Hansen, 2013).

Scholarship on extractive industries and indigenous peoples is not limited to the Canadian North. Several scholars have contributed to this body of literature over recent decades, some of whom presented nuanced positions on indigenous responses to industrial development (McNeish, 2012; O'Faircheallaigh, 2013; Stern, 2001; West, 1994). McNeish questions whether the images of indigenous resistance and environmentalism in relation to energy extraction developments in Latin America are not dangerously oversimplified (McNeish, 2012). Instead of stereotypical resistance, local indigenous communities seek dialogue to negotiate settlements that would benefit community development. O'Faircheallaigh discusses a change in approach from industry and state actors to indigenous people in western Australia that led to an increased indigenous capacity to negotiate agreements and signalled the emergence of leverage points that were not available during earlier phases of resource development (O'Faircheallaigh, 2013).

Despite these few examples of an emerging sophisticated view of indigenous agency, scholars continue to focus on community resilience, cultural impacts and risks of disturbance to the conditions for and underpinnings of cultural continuity rather than on the transformative and proactive capabilities of these communities (Wanvik & Caine, 2017). To a certain extent, indigenous people are still portrayed as responsive agents, only consulting on individual industrial development disturbances by invitation from a benevolent power holder (i.e. industry or the state) (Angell & Parkins, 2011; Chandler & Lalonde, 1998; Fleming & Ledogar, 2008). A better understanding of indigenous people's mobilization in support of their own goals and aspirations is required, including insights into how they proactively respond to industrial developments as agents in their own right, rather than focusing on their reactive adaptations.

A way of doing this is to underscore the power dynamics of contested energy spaces as more unstable and contingent (see Chap. 5). Subsequently, I argue that an analytical framework is needed to properly understand and explain recent developments among indigenous Métis communities[2] in northern Alberta. Several Indigenous communities in this region have mobilized a variety of resources to increase their leverage and expand their rights in the midst of the oil sands, counter to how the fate of Indigenous communities conventionally has been portrayed. Such transformative competence is nothing new (Pelling, 2010), but it has not been conceptually and comprehensively introduced into studies of indigenous practices. Hence, rather than being subject to circumstance, I argue that Indigenous communities seize the moment through *strategic* and *pragmatic* engagement with an ever-changing environment.

7.2   Indigenous Strategic Pragmatism

The regional municipality of Wood Buffalo is an apt case study for understanding how the dynamic between Indigenous people and extractive industry development differs from conventional explanations. This region has many of the characteristics of extractive industry developments, including a socially and economically marginalized population (O'Faircheallaigh, 2013); a history of extractive industry activity that has imposed cultural and social costs on Indigenous people but generated few benefits for them (Huseman & Short, 2012); and provincial and local government institutions and policies that have been strongly supportive of industry (Fluet & Krogman, 2009; Hanson, 2013; Huseman & Short, 2012; Justus & Simonetta, 1982; Taylor & Friedel, 2011; Wanvik, 2016; Wanvik & Caine, 2017).

Indigenous Métis groups in the region have been largely under-represented in negotiations over land use rights and the expropriation of traditional Indigenous territories (Adams, Dahl, & Peach, 2013; Andersen, 2008; Chartier, 1994, 1999; Chartrand, 2008; Sawchuk, 1985; Sosa & Keenan, 2001). Their ability to develop their own ways of engaging with state and industry has been precarious, and at the same time their strategic mobilization of resources has given them a set of properties and emergent capacities that allow them to pragmatically engage with extractive industry developments. Assemblage thinking inspired me to look for unstable and contingent relations—treating communities and institutions as emergent; constantly seeking new and productive alliances. However, Indigeneity rarely connotes innova-

---

[2] Since the inclusion of the Métis in the Canadian Constitution of 1982, they have been recognized as a distinct indigenous people The Métis originally descended from the early encounters between European settlers and First Nations.

tion and flexibility in the literature. On the contrary, evidence of continuing, fixed, traditional land use practices, language, culture and beliefs are seen as essential for most conventional identifiers of indigenous peoples (Fenelon & Hall, 2008), as are historic continuity, distinctiveness, marginalization, self-identity and self-governance (Dove, 2006). Arguably, changes to the social, cultural, economic and political northern landscape have led to changes in the relationships, practices and capabilities of northern Indigenous peoples when dealing with industry and the government[3].

According to Metis legal scholar Jean Teillet (2013), Métis indigeneity does not easily subscribe to presuppositions of fixity or reactive responses. In fact, the very concept of the Métis as a people is said to challenge the established boundaries of culture in Canada. From the beginning, the Métis have defined themselves precisely by an ability to construct and reproduce a unique but vibrant and transformative socio-economic system and culture at the interstices of Indigenous and Euro-Canadian societies (Clark, O'Connor, & Fortna, 2015), building strategic alliances across cultural and economic divides.

Strategic relations traverse history, and these relations are necessarily unstable and subject to change (Faubion, 1994; Foucault, Rabinow, & Hurley, 1997). Alliances are made, or co-functions emerge, between various component parts. Hence, the claim that power is produced or generated by strategic alliances or co-functioning distinguishes the concept of power from an inscribed capacity (Allen, 2011a, 2011b) and underpins the notion of the emergent capacities of component parts (DeLanda, 2006, 2016). All of these strategic relationships are local, regional forms of power that have their own ways of functioning and their own procedures and techniques. Therefore, we cannot speak of power. Instead, we must speak of powers, and try to localize them in their historical and geographical specificity. I suggest that such a conceptualization of power is suggestive of a certain form of pragmatism (O. Jones, 2008). Co-functioning, or the pooling of resources in specific events or situations, produces power. Component parts may occasionally align, or pool their resources with other component parts, as a means of securing common goals (Allen, 1997, 2003, 2011a, 2011b). Drawing on Foucault, I claim that the Wood Buffalo contested energy space is an archipelago of different powers (Foucault, 2007). The contested energy space is not a unitary body in which one power exercises itself; in reality, it is a situation; a liaising and a co-ordination of different powers and interests that nonetheless retain their specificity.

Situations create a variety of interests (or stakes) among the component parts (or stakeholders) of the assemblage. Thus, component parts may be defined as stakeholders in a given situation. Although stakeholder theory primarily focuses on the management of companies and their operative environments (Carroll & Buchholtz, 2014; Fassin, 2009; Freeman, 2010), we argue for a broader scope in which a stakeholder framework can be used in concert with assemblage theory in order to shed some light on stakeholder relationships within particular situations, such as the extractive energy landscapes of Alberta, where different component parts have different interests or

---

[3] Amendments to the Indian Act, which now allows First Nations communities to take legal action, and the subsequent important role of court cases have strongly contributed to these changes.

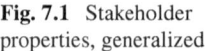

Fig. 7.1 Stakeholder
properties, generalized

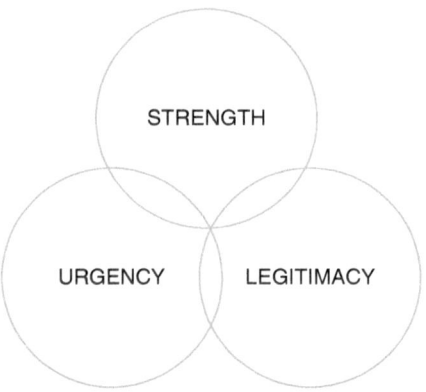

stakes in the production of the contested energy spaces in Wood Buffalo. However, the ability to defend these interests varies greatly between stakeholders. Based on the grounded data analysis, I have identified three stakeholder properties: (1) strength, (2) urgency and (3) legitimacy. Stakeholder strength is closely related to the number of members and the amount of funds or other types of resources that amplify the stakeholder's position. Stakeholder urgency underscores how important the time factor is and how critical the stakeholder's expectations are. Stakeholder legitimacy refers to the status of the stakeholder's claim or interest, which is most commonly related to consultative status for Indigenous communities. Figure 7.1 shows a standardized stakeholder with all properties equally high (strength, urgency and legitimacy = high), signifying an ultimate stakeholder (Mitchell, Agle, & Wood, 1997). All of the properties are relationally defined, which indicates that the value of each property could depend upon the situation in which it is used.

By approaching Indigenous communities as goal-motivated, pragmatic component parts of the contested energy space of Wood Buffalo, I seek to energize the conceptual frameworks of assemblage thinking with a more practical vocabulary of strategic pragmatism. Hence, I refer to processes that work to stabilize or destabilize the assemblage as the *mobilization of resources,* wherein component parts align to generate power to change or uphold the assemblage. I refer to the properties gained by component parts and their mobilization as *outputs,* underscoring the short-term potential character of properties if they are not activated. I refer to the capacities that emerge from the outputs as *outcomes,* to acknowledge the more long-term emergent effects of the co-functioning and aligning of component parts. The longer term fundamental changes in the assemblage composition that derive from these processes I term *impacts* (Wanvik & Caine, 2017).

7.3   Employing Strategic Pragmatism

Change is something Métis (Elder, McMurray Métis, interview 2016).

Indigenous Métis communities in this study regard themselves as rather small, with fewer than 300 members in each community (strength = low). Moreover, even

Fig. 7.2 Métis properties

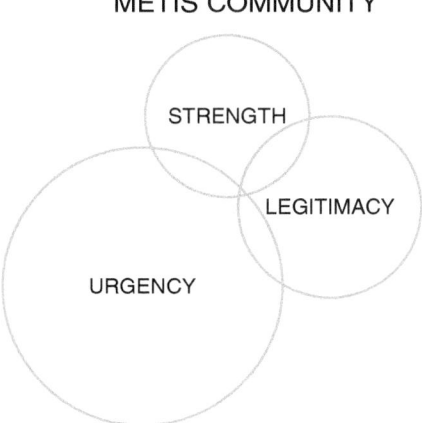

METIS COMMUNITY

STRENGTH

LEGITIMACY

URGENCY

in areas where they have come into direct contact with extractive activities, they consider themselves to have a low degree of de facto legitimacy (legitimacy = low) because the Alberta government has not granted them consultative status as indigenous people. However, they most certainly claim legitimacy as right-bearing communities, and there is a strong degree of urgency (urgency = high) to their claims due to the threat to traditional land use practices posed by industrial developments. Thus, as we will see, to strengthen their position as stakeholders, Indigenous Métis communities strategically mobilize resources (legitimacy and strength) by engaging other stakeholders (Fig. 7.2).

Mobilizing Legitimacy and Strength Through Coalitions

It has been critical for Métis communities to manoeuvre into a negotiating position. There has been no easy path to negotiating with industry representatives without formal political consultative status for Métis communities; therefore, they have had to explore other pragmatic strategies. For the Fort McKay Métis, the aligning of interests between the Métis and First Nation communities was the first step in a mutual effort to sign privately negotiated community–industry impact benefits agreements (IBAs).

When we started working with industry, we were the first Metis community to sign an impact benefits agreement. The reason we managed to do it was because we piggybacked the First Nation. The First Nation included us in the process to gain a larger head count, which enabled us to sign an agreement (President, Fort McKay Métis, interview 2015).

The First Nation community of Fort McKay is adjacent to the Métis community, there is a close kinship between the two groups, and to some extent, they share parts of each other's traditional lands: "They needed us and we needed them" (CEO, Fort McKay Métis, interview 2016). First Nations already have consultative status accord-

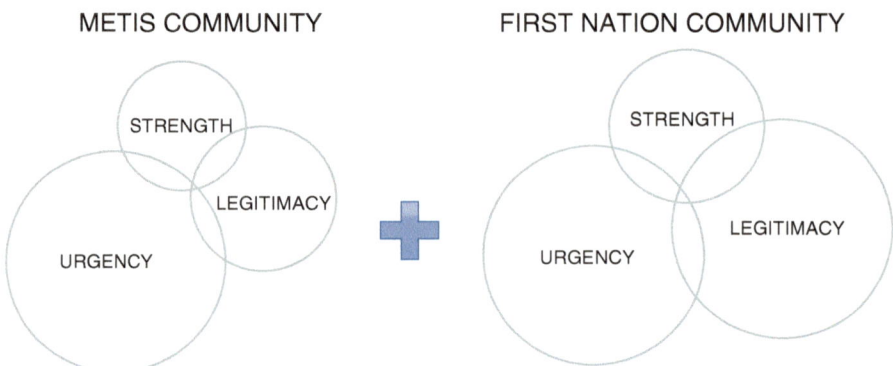

Fig. 7.3 Metis and First Nation alliance

ing to the Alberta consultation policy (Government of Alberta, 2014). By strategically joining the First Nation in their negotiations with industry, the Fort McKay Métis gained sufficient legitimacy to participate, even as an independent community.

First Nations represented the Métis, and did a number of studies, probably 30–50 studies, proving that both communities overlap. We have been able to use that as leverage (President, Fort McKay, interview 2015).

Today, the Métis continue to negotiate on their own, albeit without formal consultative status, proving that, to some extent, precedents established their emergent consultative capacity: "We have negotiated four, soon to be five, agreements on our own now. They all are long-term, 40-year agreements" (President, Fort McKay Métis, interview 2016). Using empirical findings to illustrate the mobilizing of legitimacy and strength by aligning with the neighbouring First Nation community, Fig. 7.3 shows the relationship between the two component parts and their self-proclaimed properties.

According to informants, the indigenous groups compensated for both groups' deficiency in numbers (strength = low/medium) by joining forces. What the Métis community lacked in consultative status (legitimacy = low) and negotiating capacity (strength = low), they gained by aligning with a group that has such status. By engaging First Nation properties, the Métis gain an emergent capacity of having a negotiating position with industry on their own.

Mobilizing Legitimacy and Strength Through External Competence

My informants acknowledged that local communities are more capable of "playing" the different stakeholders now than they were before. In particular, hiring external consultants to facilitate and accommodate organizational development and advocacy work has contributed to capacity development and the expansion of social and symbolic capital (Caine, 2016): "We hire the best person for the job, and I don't care if he or she is white or black or even Muslim" (Elder, McMurray Métis, interview 2016). All three of the communities in this study have used industry funding to hire external management staff with competence in organizational development, negotiations and public policy. External competence has become part and parcel of the organizational processes of the three Métis communities. There is also a scalar dimension to the hiring of external competence. With these resource persons, the Métis communities even gained access to decision makers at the federal level:

> The previous CEO [externally hired] was good with contracts, you see? I told him, there is a good chance you will go to Ottawa [federal government], and I need a contact there. He got a job there now, and so far, he set us up with two ministers now, Indian and northern affairs, and infrastructure. He has paid off long time ago. Even the prime minister came to see us. I have never seen anybody like him, man. He pushed me, you know […]; he told me to marry politics and economics to get things done, eh? (Elder, McMurray Métis, interview 2016).

This informant echoes the importance of external competence regarding his own personal development, but he also states how external experts remain assets for the communities after they have moved on. To illustrate the mobilizing of legitimacy and the strength acquired by hiring external experts and consultants, Fig. 7.4 shows the proposed relationship between the two component parts and their properties.

According to the informants, the lack of strength and legitimacy among Métis communities (= low) was compensated for by the consultants' specific competencies in advocacy, organizational development and negotiating skills, in addition to their networks and contacts with decision makers at local, regional or federal levels.

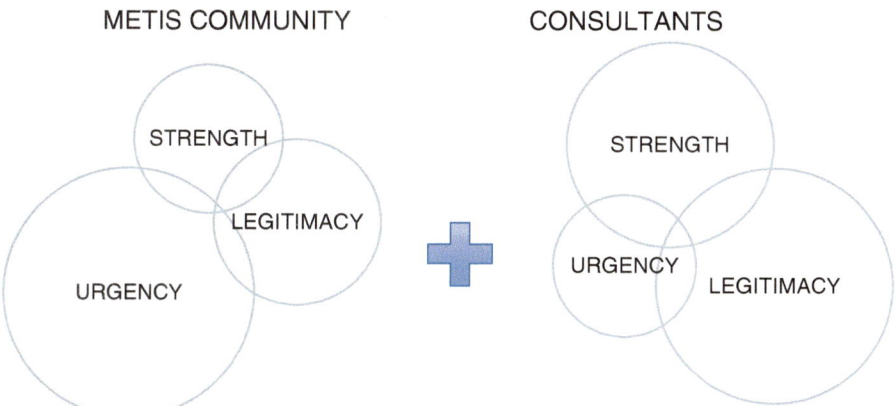

Fig. 7.4 Métis and consultants agreements

Mobilizing Strength Through Industry Agreements

From the Conklin Métis I learned of another kind of aligning of interests that under-scores the benefits of shared interests between local indigenous communities and the extractive industry:

> The indigenous groups are more empowered now than ever, more vocal than ever, and more involved. When they [Indigenous groups] are not being recognized, all of that comes into the mix when industry tries to operate. (CSR manager, Statoil Canada, interview 2015).

Corporate motivation for engaging with local indigenous communities can be divided into three categories: (1) managing risk, (2) creating a loyal working envi-ronment and (3) developing a pool of skilled local labour (Wanvik, 2016). Legal consultants working with the local communities claim that:

> Most indigenous Métis communities recognize that the regulatory process is biased in favour of extractive industry development, and communities seek economic and contracting benefits because the choices they are faced with are either having development proceed and receiving some benefit from it, or having development proceed and receiving no benefit at all from it (Community lawyer, email correspondence, 2015).

This statement is supported by our findings from the Métis community leaders:

> You can fight the oil sands and get nothing, or you can play with them, and get something back. Industry is all the same; it is either their way or the highway. But they want to play with us because they know that we can stop industry with our rights (Elder, McMurray Métis, interview 2016).

Taken together, our findings point to a shared understanding of the need for a pragmatic alignment of interests between communities and companies. To illustrate this, Fig. 7.5 shows the relationship between the two component parts and their properties: for the Métis, companies are first and foremost approached for achieving

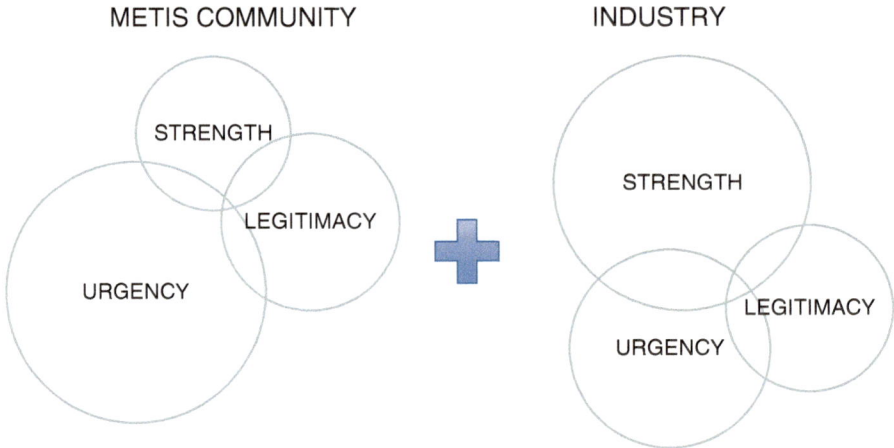

Fig. 7.5 Métis and industry agreements

financial support (strength = high), which enables them to increase their leverage, support local development and welfare, and to document their traditional territory and land use practices (as we shall see in the next section).

In short, the territorializing processes of mobilizing resources to strengthen the Métis communities are important factors in the relational game of contested energy spaces. Different component parts engage each other, negotiate and find common ground to move forward. In the following, we look at the realization of short-term properties (outputs) and how these properties are invested in real and envisioned outcomes to produce emerging capacities for change and improvements in community well-being and political leverage.

Outputs: Properties Gained Through Impact Benefits Agreements

The major output of the strategic mobilization of resources has been the formalization of agreements between the Métis communities and different extractive companies. As we have seen (previous chapter), these agreements are commonly referred to as IBAs, which are privately negotiated agreements, typically between extractive industries and local communities, whereby the government is relegated to an external, observational role. IBAs establish formal relationships between signatories, mitigating negative development impacts and enhancing positive development outcomes for indigenous communities (O'Faircheallaigh, 1999, 2007; Wanvik, 2016):

> We used to say, 'you have a project, give us a business contract, and we let you do your project'. We did not look after the bigger picture. So we started working with IBAs. We were the first Métis community to sign an IBA (President Fort McKay Métis, interview 2016)

These agreements have primarily focused on properties such as local employment and economic benefits (Statoil Canada, 2007), whereas more recent IBA constructions acknowledge the need for greater flexibility and diversity of community involvement in industrial decisions, as well as the need for social and cultural programs, dispute resolution mechanisms, revenue-sharing provisions and environmental restrictions (Caine & Krogman, 2010; Dreyer & Myers, 2005; Gibson, 2008; Sosa & Keenan, 2001). These agreements have provided communities with direct funding support for physical, social and human infrastructure, as well as contracting opportunities for company businesses and a process to address environmental issues involving future developments (Wanvik, 2016).

Resource development proponents have an incentive to enter into IBAs with indigenous communities to obtain indigenous communities' consent for access to their lands for resource development, obtain labour locally and create a co-operative working relationship (Caine & Krogman, 2010; Wanvik, 2016). However, there are challenges related to the sustainability of these agreements. Industrial partners may not have the resources and competencies to actively maintain local development over time, and it is not clear what happens when investments dry up and industry leaves (Wanvik, 2016). Other mediated challenges concern the region's heavy dependence on the oil industry:

> Diversification is really important to us now, with the downturn. We are not putting all our eggs in one basket in terms of serving the oil sands. We are looking at tourism, foods and beverages, and so on. Sustainability is everything. […] If I leave, and the community cannot sustain itself, for me that is a failure (President, Fort McKay Métis, interview 2016).

IBAs have enabled the Métis communities in many ways as business partners and subcontractors, as historic, rights-bearing communities, and as educational institutions for their own youth population. However, it is hoped that the main outcome of these IBAs is that they outlive the industrial developments of the Wood Buffalo energy bonanza, which leads us to traditional land use studies (TLUs).

Outcomes: Capacities Activated Through Traditional Land Use Studies

Among the outcomes of IBAs are TLUs, which may be the most influential factor related to Indigenous community capacity building. Indigenous communities use TLUs as territorializing tools to identify and assess their traditional land use practices to claim compensation for or exert influence on extractive industry developments. All three of the studied Métis communities have made comprehensive TLUs (some of them have made several) in the course of industry negotiations (Clark et al., 2015; Conklin Métis Local #193, 2012; Fort McKay Sustainability Centre (TLU), 2016):

> We are still struggling to get acknowledgement as a right-based community. Thank God, we did our homework. Industry paid for it. We needed that. I got financial support to record my elders, so in the early 2000s, we got their stories (Clark et al., 2015) […]. That was the one that made a change (Elder, McMurray Métis, interview 2016).

The importance of TLUs can hardly be overestimated among our informants:

> Suddenly, we have our land use practices in writing, on the coffee tables of decision makers and industry, where they can read about our territory. All of a sudden, there is that truth to it, yeah? Do people read it? Sure! They got no choice now, right? (Elder, McMurray Métis, interview 2016).

Industry seems to acknowledge the implications of their support for indigenous communities:

> We help fund traditional land use studies (referred to as Conklin Métis Local #193, 2012) and cultural initiatives. They have been asking for this for decades, but eventually, it is the industry that is providing the studies (CSR manager, Statoil Canada, interview 2015).

These outcomes are generating increased capacity building among the Métis communities, underscoring the territorializing function of these TLUs and the direct impact they have on negotiations with industry. Together with competence building related to negotiations, these activities are all part of the new reality of indigenous communities. Hence, the communities themselves are calibrating their participatory role in the emerging governance processes of the contested energy space of the Wood Buffalo to strengthen Métis negotiating power.

Impacts: Alliance Building and Re-Territorializing Processes

The strategic mobilization of resources, the formalization of IBAs and the subsequent increased focus on TLUs to document historic practices have all contributed to substantial collaboration between the Métis communities of Wood Buffalo. In 2015, the Métis communities of Conklin, Fort McMurray, Fort McKay and Anzac formed the Wood Buffalo Métis. Because each community is too small on its own, the Métis communities need to work together to have their voices heard. By sharing competence through the exchange of external consultants and hired experts, and by reactivating and strengthening historic bonds through the TLUs, the Métis communities of Wood Buffalo grew closer together:

> Because of the One Nation view of the Métis, we are not like many First Nations, you know, like a nation here, a nation there. The Métis think a little differently, right? We are communities within a bigger community, a Métis Nation, so we are nationalists to that extent, so to speak, so it is a natural for us to think that we are part of the same community (CEO, Fort McKay Métis, interview 2016).

This notion feeds into the narrative of Métis collaboration across communities, with one common goal:

> Together, we have made many moves, all through teamwork. When we are recognized as a rights-bearing community, much will have to change (Elder, McMurray Métis, interview 2016).

7.4   Recent Developments

Claiming Land Through Purchase

In March 2018, a $1.6-million deal was brokered between Fort McKay Métis and the Alberta government. The community took ownership of 150 hectares of land it had currently been leasing since 2007. In 2014 the Métis community bought almost 49 hectares for $1 to use for housing. By 2018, the deal for the remaining 150 hectares was complete (Clancy, 2018).

It is believed to be the first Metis settlement to buy all the land it occupies from a provincial government, and the community, which has members throughout Canada, is now calling on people to "come home" (President Quintal, quoted in Clancy, 2018). "It was a much quicker process than having to continually negotiate," he said, adding changes in government ministries over the past decade slowed the process and spurred the community to work with a bank, said Quintal. "We're breaking ground on a new community pavilion," he said. Cultural and healing centres are also in the works along with a gas station, car wash and Tim Hortons. The goal is to deliver services ranging from grocery stores to barber shops within the community, he said (Graveland, 2018).

United on Pipelines?

First Nations and Métis communities in the Fort McMurray region are expressing interest in becoming business partners in the pipeline industry. The indigenous communities want to either buy a stake in the Trans Mountain pipeline or partner and build another future line. "We want to be owners of a pipeline," Allan Adam, chief of Athabasca Chipewyan First Nation, said in an interview (Thurton, 2018). "We think that a pipeline is a critical component to the oil and gas sector, especially in this region." Quintal said he expects they would need backers to help guarantee loans to help fund the multi-billion-dollar project. Participants say it was the first time the region's Cree, Dene and Métis communities met together with the head of the federal government. The announcement from Adam is a change in position for the chief, who is no stranger to pipeline opposition. The chief has posed with celebrities and activists critical of the oil sands' environmental legacy (De Souza, 2016). "The fact is I am tired. I am tired of fighting. We have accomplished what we have accomplished," Adam said. "Now let's move on and let's start building a pipeline and start moving the oil that's here already" (De Souza, 2016). Also, he said, Indigenous owners would take the utmost care to ensure the pipeline route would avoid sacred or sensitive land areas and that infrastructure is maintained to the highest standards to prevent spills. Adams' Métis partner claimed that this development was not meant as an open book or blank cheque for the industry to develop pipelines. "Ultimately we are the keepers of the land and it is of the utmost importance that lands are protected as much as possible". Quintal said Indigenous owners behind a pipeline might also lend credibility to a project that could quell some opposition (De Souza, 2016). However, Adam said that the idea is still preliminary and that he and other chiefs still have to take the announcement back to their respective First Nations to seek out their approval (Toth, 2018). There he might find quite some opposition, both from elders and young activists.

7.5 Conclusion

We witness the evolution of a truly strategic and pragmatic political agency comprising several indigenous Métis communities capable of mobilizing resources to increase their influence and strengthen their rights vis-à-vis government institutions and industry in the contested energy spaces of Wood Buffalo. They gain financial support by negotiating IBAs; they invest in TLUs and make decisions about the procurement of additional competencies and the hiring of management staff and they capitalize on these capacities by creating new and more powerful coalitions and alliances.

References

Adams, C., Dahl, G., & Peach, I. (2013). *Métis in Canada: History, identity, law and politics*. Edmonton, Canada: University of Alberta.

Allen, J. (1997). Economics of power and space. In R. Lee & J. Wills (Eds.), *Geography of economies*. London, UK: Arnold.

Allen, J. (2003). *Lost geographies of power*. Oxford, UK: Blackwell Publishing.

Allen, J. (2011a). Powerful assemblages? *Area, 43*(2), 154–157. https://doi.org/10.1111/j.1475-4762.2011.01005.x

Allen, J. (2011b). Topological twists: Power's shifting geographies. *Dialogues in Human Geography, 1*(3), 283–298. https://doi.org/10.1177/2043820611421546

Andersen, C. (2008). From nation to population: The racialisation of 'Métis' in the Canadian census. *Nations and Nationalism, 14*(2), 347–368.

Angell, A. C., & Parkins, J. R. (2011). Resource development and aboriginal culture in the Canadian north. *Polar Record, 47*(1), 67–79. https://doi.org/10.1017/S0032247410000124

Arena, C., Bozzolan, S., & Michelon, G. (2015). Environmental reporting: Transparency to stakeholders or stakeholder manipulation? An analysis of disclosure tone and the role of the board of directors. *Corporate Social Responsibility and Environmental Management, 22*(6), 346–361. https://doi.org/10.1002/csr.1350

Barker, A. J. (2015). 'A direct act of resurgence, a direct act of sovereignty': Reflections on idle no more, Indigenous activism, and Canadian settler colonialism. *Globalizations, 12*(1), 43–65.

Caine, K. J. (2016). Blurring the boundaries of environmentalism: The role of Canadian Parks and Wilderness Society as a boundary organization in northern conservation planning. *Rural Sociology, 81*(2), 194–223.

Caine, K. J., & Krogman, N. (2010). Powerful or just plain power-full? A power analysis of impact and benefit agreements in Canada's north. *Organization & Environment, 23*(1), 76–98. https://doi.org/10.1177/1086026609358969

Carroll, A. B., & Buchholtz, A. K. (2014). *Business and society: Ethics, sustainability, and stakeholder management*. Stamford, CT: Nelson Education.

Chandler, M. J., & Lalonde, C. (1998). Cultural continuity as a hedge against suicide in Canada's First Nations. *Transcultural Psychiatry, 35*(2), 191–219.

Chartier, C. (1994). *Métis perspective*. Paper presented at the Continuing Poundmaker and Riel's Quest: Presentations Made at a Conference on Aboriginal Peoples and Justice.

Chartier, C. (1999). Aboriginal self-government and the Metis nation. *Aboriginal Self-Government in Canada: Current Trends and Issues, 2*, 112–128.

Chartrand, L. (2008). 'We rise again': Metis traditional governance and the claim to Metis self-government. In Y. Belanger (Ed.), *Aboriginal self-government in Canada* (3rd ed., pp. 145–157). Saskatoon, Canada: Purich Publishing.

Clancy, C. (2018, March 28). "We're trailblazers": Fort McKay Métis community celebrates $1.6-million land deal, News. *Edmonton Journal*. Retrieved from http://edmontonjournal.com/news/local-news/were-trailblazers-fort-mckay-metis-community-celebrates-1-6-million-land-deal

Clark, T. D., O'Connor, D., & Fortna, P. (2015). *Mark of the Metis - Fort McMurray: Historic and contemporary rights-bearing Métis community*. Retrieved from Alberta, Canada.

Coates, K. (2015). *# IdleNoMore: And the remaking of Canada*. Saskatchewan, Canada: University of Regina Press.

Conklin Métis Local #193. (2012). *Stories from a long time ago*. Retrieved from Conklin, Alberta, Canada.

De Souza, M. (2016, December 13). First Nations chief at heart of oilsands resistance says he's no environmentalist, News. *National Observer*. Retrieved from https://www.nationalobserver.com/2016/12/13/news/first-nations-chief-heart-oilsands-resistance-says-hes-no-environmentalist

DeLanda, M. (2006). *A new philosophy of society: Assemblage theory and social complexity*. London, UK: A&C Black.

DeLanda, M. (2016). *Assemblage theory*. Edinburgh, Scotland: Edinburgh University Press.

Dove, M. R. (2006). Indigenous people and environmental politics. *Annual Review of Anthropology, 35*(1), 191–208.

Dreyer, D., & Myers, H. (2005). *Impact and benefits agreements: Do the Ross River Dena benefit from mineral projects?* Prince George, Canada: University of Northern British Columbia.

Erikson, K. T. (1976). *Everything in its path*. New York, NY: Simon and Schuster.

Fassin, Y. (2009). The stakeholder model refined. *Journal of Business Ethics, 84*(1), 113–135.

Faubion, J. D. (1994). *Essential works of Foucault 1954-1984, vol 3: Power*. London, UK: Penguin.

Fenelon, J. V., & Hall, T. D. (2008). Revitalization and indigenous resistance to globalization and neoliberalism. *American Behavioral Scientist, 51*(12), 1867–1901.

Fidler, C. (2010). Increasing the sustainability of a resource development: Aboriginal engagement and negotiated agreements. *Environment, Development and Sustainability, 12*(2), 233–244. https://doi.org/10.1007/s10668-009-9191-6

Fleming, J., & Ledogar, R. J. (2008). Resilience, an evolving concept: A review of literature relevant to aboriginal research. *Pimatisiwin, 6*(2), 7.

Fluet, C., & Krogman, N. (2009). The limits of integrated resource management for Alberta for aboriginal and environmental groups: The Northeast Slopes sustainable resource and environmental management strategy. In *Environmental conflict and democracy in Canada* (pp. 123–139). Vancouver, Canada: UBC Press.

Fort McKay Sustainability Centre (TLU). (2016). *Teck Frontier Mine Project - Fort McKay Métis Integrated Cultural Assessment*. Retrieved from Fort McMurray, Canada.

Foucault, M. (2007). The meshes of power. In *Space, knowledge and power: Foucault and geography* (pp. 153–162). Aldershot, UK: Ashgate.

Foucault, M., Rabinow, P., & Hurley, R. (1997). *The essential works of Michel Foucault, 1954-1984*. London, UK: Penguin.

Freeman, R. E. (2010). *Strategic management: A stakeholder approach*. New York, NY: Cambridge University Press.

Gibson, G., & Klinck, J. (2005). Canada's resilient north: The impact of mining on aboriginal communities. *Pimatisiwin, 3*(1), 116–139.

Gibson, V. V. (2008). *Negotiated spaces: Work, home and relationships in the Dene diamond economy*. Vancouver, Canada: University of British Columbia.

Government of Alberta. (2014). *Government of Alberta proponent's guide to First Nations consultation procedures for land dispositions*. Retrieved from Calgary.

Graveland, B. (2018, March 28). Small northern Alberta community gets bigger with historic land purchase, News story. *Global News*. Retrieved from https://globalnews.ca/news/4111647/fort-mckay-metis-land-purchase-alberta/

Hanson, L. L. (2013). Changes in the social imaginings of the landscape. The management of Alberta's rural public lands. In J. R. Parkins & M. G. Reed (Eds.), *Social transformation in rural Canada: New insights into community, cultures, and collective action* (pp. 148–168). Vancouver, Canada: UBC Press.

Harvey, B., & Bice, S. (2014). Social impact assessment, social development programmes and social licence to operate: Tensions and contradictions in intent and practice in the extractive sector. *Impact Assessment and Project Appraisal, 32*(4), 327–335. https://doi.org/10.1080/14615517.2014.950123

Healey, S. (2006). *Cultural resilience, identity and the restructuring of political power in Bolivia.* Paper presented at the Paper submitted for the Eleventh Biennial Conference of the International Association for the Study of Common Property. Bali, Indonesia: Conference of the International Association for the Study of Common Property.

Hoberg, G., & Phillips, J. (2011). Playing defence: Early responses to conflict expansion in the oil sands policy subsystem. *Canadian Journal of Political Science, 44*(3), 507–527.

Hovelsrud, G. K., & Krupnik, I. (2006). IPY 2007-08 and social/human sciences: An update. *Arctic, 59*(3), 341–348.

Huseman, J., & Short, D. (2012). 'A slow industrial genocide': Tar sands and the indigenous peoples of northern Alberta. *The International Journal of Human Rights, 16*(1), 216–237.

Jones, O. (2008). Stepping from the wreckage: Geography, pragmatism and anti-representational theory. *Geoforum, 39*(4), 1600–1612. https://doi.org/10.1016/j.geoforum.2007.10.003

Justus, R., & Simonetta, J. (1979). *Major resource impact evaluation.* Cold Lake, Canada: Cold Lake Band and Indian and Northern Affairs Canada.

Justus, R., & Simonetta, J. (1982). Oil sands, Indians and SIA in northern Alberta. In *Indian SIA: The social impact assessment of rapid resource development on Native Peoples* (pp. 238–257). Ann Arbor, MI: University of Michigan Press.

Kirmayer, L. J., Brass, G. M., & Tait, C. L. (2000). The mental health of Aboriginal peoples: Transformations of identity and community. *The Canadian Journal of Psychiatry, 45*(7), 607–616.

Lawrence, S., & Macklem, P. (2000). From consultation to reconciliation: Aboriginal rights and the Crown's duty to consult. *Canadian Bar Review, 29*, 252–279.

Le Billon, P., & Carter, A. (2012). Securing Alberta's tar sands: Resistance and criminalization on a new energy frontier. In *Natural resources and social conflict: Towards critical environmental security* (pp. 170–192). Basingstoke, UK: Palgrave Macmillan.

Magis, K. (2010). Community resilience: An indicator of social sustainability. *Society and Natural Resources, 23*(5), 401–416.

McMurray Metis (TLU). (2015). *Fort McMurray: Historic and contemporary rights-bearing Métis community.* Retrieved from Alberta, Canada.

McNeish, J.-A. (2012). More than beads and feathers: Resource extraction and the indigenous challenge in Latin America. In *New political spaces in Latin American natural resource governance* (pp. 39–60). New York, NY: Palgrave Macmillan.

Mitchell, R. K., Agle, B. R., & Wood, D. J. (1997). Toward a theory of stakeholder identification and salience: Defining the principle of who and what really counts. *Academy of Management Review, 22*(4), 853–886.

O'Faircheallaigh, C. (1999). Making social impact assessment count: A negotiation-based approach for indigenous peoples. *Society & Natural Resources, 12*(1), 63–80.

O'Faircheallaigh, C. (2007). Environmental agreements, EIA follow-up and aboriginal participation in environmental management: The Canadian experience. *Environmental Impact Assessment Review, 27*(4), 319–342. https://doi.org/10.1016/j.eiar.2006.12.002

O'Faircheallaigh, C. (2010a). Aboriginal-mining company contractual agreements in Australia and Canada: Implications for political autonomy and community development. *Canadian Journal of Development Studies/Revue canadienne d'études du développement, 30*(1-2), 69–86. https://doi.org/10.1080/02255189.2010.9669282

O'Faircheallaigh, C. (2010b). Public participation and environmental impact assessment: Purposes, implications, and lessons for public policy making. *Environmental Impact Assessment Review, 30*(1), 19–27. https://doi.org/10.1016/j.eiar.2009.05.001

O'Faircheallaigh, C. (2013). Extractive industries and Indigenous peoples: A changing dynamic. *Journal of Rural Studies, 30*, 20–30.

Pelling, M. (2010). *Adaptation to climate change: From resilience to transformation.* London, UK: Routledge.

Prno, J., & Scott Slocombe, D. (2012). Exploring the origins of 'social license to operate' in the mining sector: Perspectives from governance and sustainability theories. *Resources Policy, 37*(3), 346–357. https://doi.org/10.1016/j.resourpol.2012.04.002

Sawchuk, J. (1985). The Metis, non-status Indians and the new Aboriginality: Government influence on native political alliances and identity. *Canadian Ethnic Studies= Etudes Ethniques au
Canada, 17*(2), 135.

Sosa, I., & Keenan, K. (2001). *Impact benefit agreements between aboriginal communities and
mining companies: Their use in Canada.* Toronto, Canada: Canadian Environmental Law
Association.

Statoil Canada. (2007). *Application for approval of the Kai Kos Dehseh Project.* Retrieved from
Calgary, Canada.

Stern, P. R. (2001). *Modernity at work: Wage labor, unemployment, and the moral economy of
work in a Canadian Inuit community.* Ann Arbor, MI: University Microfilms International.

Taylor, A., & Friedel, T. (2011). Enduring neoliberalism in Alberta's oil sands: The troubling
effects of private–public partnerships for First Nation and Métis communities. *Citizenship
Studies, 15*(6-7), 815–835.

Teillet, J. (2013). *Métis law in Canada* (0991702700). Retrieved from Alberta, Canada. http://
albertametis.com/wp-content/uploads/2014/04/Metis-Law-in-Canada-2013-1.pdf

Thurton, D. (2018, April 15). 'We want to be owners': Fort McMurray First Nations and Métis
unite on pipelines, News. *CBC.* Retrieved from http://www.cbc.ca/news/canada/edmonton/
fort-mcmurray-first-nations-and-m%C3%A9tis-pipelines-1.4613003

Toth, K. (2018, April 16). 'It hurt': First Nations leaders' pipeline ownership proposal comes
as shock to some, News. *CBC.* Retrieved from http://www.cbc.ca/news/canada/north/
fort-chip-pipeline-proposal-1.4621885

Usher, P. J. (2000). Traditional ecological knowledge in environmental assessment and management. *Arctic, 53,* 183–193.

Waldrum, J. B. (1988). *As long as the rivers run: Hydroelectric development and native communities in western Canada.* Winnipeg, Canada: University of Manitoba Press.

Wanvik, T. I. (2016). Governance transformed into CSR – new governance innovations in the
Canadian oil sands. *The Extractive Industries and Society, 3,* 517.

Wanvik, T. I., & Caine, K. (2017). Understanding indigenous strategic pragmatism: Métis engagement with extractive industry developments in the Canadian North. *The Extractive Industries
and Society, 4*(3), 595–605. https://doi.org/10.1016/j.exis.2017.04.002

West, P. C. (1994). Natural resources and the persistence of rural poverty in America: A Weberian
perspective on the role of power, domination, and natural resource bureaucracy. *Society &
Natural Resources, 7*(5), 415–427.

Wotherspoon, T., & Hansen, J. (2013). The "Idle No More" movement: Paradoxes of First Nations
Inclusion in the Canadian context. *Social Inclusion, 1*(1), 21.

Chapter 8
Case Study III: The Role of Non-Human Agency

Abstract This chapter investigates the impact of non-human agency within the contested energy spaces of the Canadian North. By analysing the aftermath of "The Beast"—the devastating Fort McMurray wildfire of 2016—the author gives an account of the great impact natural events might have on societal relations, triggering developments in the relationship between government and Indigenous communities that in the case of Wood Buffalo have been crucial, but subject to inertia for decades.

Keywords Non-human agency · Wildfire · Disaster politics · Assemblage · Canada · Indigenous · Governance · Mobilizing

In May 2016, a massive wildfire, locally named "the Beast", forced the evacuation of close to 90,000 people from Fort McMurray and the surrounding area. Soon the flames had destroyed 2400 homes, caused the shutdown of two key oil sands facilities (Alary, 2016) and burned almost 5900 km^2 of forest. While these are all significant impacts, the aftermath of the fire reveals political consequences for local communities in the oil sands that might far outpace the material damages. Shortly after the wildfire had been contained, regional municipality representatives gathered to take stock and advise on the recovery and rebuilding of the impacted areas of Fort McMurray. The elected officials decided to suspend close to 100 rural development projects in order to ensure financial support for the recovery. This deferral was a major blow to local Indigenous communities in dire need of improved service infrastructure and other public services. It resulted in a massive mobilization among indigenous peoples in Wood Buffalo in an act of solidarity. In the weeks and months that followed, Indigenous peoples saw an opportunity to use the rebuild after the wildfire to reset the relationship between the urban and rural areas. One particular alliance, the Wood Buffalo Métis, consisting of several Métis communities from Fort Chipewyan, Fort McMurray, Fort McKay and Conklin[1], was central in the rural

---

[1] Métis Local #63 Fort McKay, Métis Local, #125 Fort Chipewyan, Métis Local #193 Conklin, and, Métis Local #1935 Fort McMurray.

© The Author(s) 2018, under exclusive licence to Springer International Publishing AG, part of Springer Nature 2019
T. I. Wanvik, *Contested Energy Spaces*, SpringerBriefs in Geography, https://doi.org/10.1007/978-3-030-02396-6_8

mobilization, and played a vital part in the political negotiations in the aftermath of the fire.

There have been considerable scholarly contributions concerning the workings of power relations within contested energy spaces. The literature often focuses on resource conflicts and environmental impacts on Indigenous peoples' traditional territories (Bebbington, 2012; Bebbington, Hinojosa, Bebbington, Burneo, & Warnaars, 2008; Black, D'Arcy, & Weis, 2014; Haarstad & Fløysand, 2007; Huseman & Short, 2012; Le Billon & Carter, 2012; LeClerc & Keeling, 2015). Characteristically, the conflicts emerging out of localized energy production are portrayed as skewed in favour of powerful stakeholders like the state or the industry, and Indigenous efforts to limit massive exploitation and subsequent destruction of their lands are portrayed as resistance or subversive activities.

The contested energy spaces of Alberta, Canada make up a truly remarkable landscape, elsewhere categorized as part of a global carbonscape (Haarstad & Wanvik, 2016). As I have argued, there are obvious structures of inertia and permanence in these energy assemblages, akin to permanent power relations of winners and losers (Wanvik & Caine, 2017). This notion of permanent power relations is amplified recently by the global economic downturn, particularly within the energy sector, followed by terrifying natural disasters, like the most recent Fort McMurray wildfire. Economic benefits and infrastructure for Indigenous communities are threatened after these devastating events, either due to cancelling of industrial developments and social investments, or recently due to reallocation of municipal funds to wildfire recovery.

Nevertheless, throughout this book I have argued that there are no such things as permanent power structures within the contested energy spaces. Instead, these assemblages are characterized by rupture, unpredictability and instability (DeLanda, 2016; Haarstad & Wanvik, 2016; Wanvik, 2014): in terms of the landscape itself, but also regarding its governance practices and power distribution. In order to understand these developments, I develop a theoretical framework fusing disaster politics analysis (Pelling & Dill, 2009) and multidimensional assemblage theory, where I argue that crisis and natural disasters can be seen to act as transformative, non-human catalysts, or assemblage converters as I label them, subsequently creating destabilizing processes within the contested energy spaces, where assemblage component parts renegotiate their relationships, connections and structures.

8.1 Disaster and Crisis as Catalytic Assemblage Converter

Geographers have long asserted a politics of disaster, mainly in relation to development and vulnerability studies (Albala-Bertrand, 1993; Birkmann, 2006; Özerdem & Jacoby, 2006). Analysis of disaster politics focuses on the interaction of social and political actors and framing institutions in preparing for and responding to extreme natural events, and suggests that the disaster events and their management are part of unfolding political histories (Hewitt, 2014). Anecdotal evidence suggests that the socio-political and cultural dynamics put into motion at the time of cata-strophic "natural" disasters create the conditions for potential political change—often at the hands of a discontented civil society (Pelling & Dill, 2006). In 1983, Michael Watts argued from a materialist perspective that disasters both emerged from pre-existing social relations and had the power to catalyse or trigger further change (Watts, 1983). Further, Albala-Bertrand argues that responses to disasters tend to reveal dominant political philosophies (Albala-Bertrand, 1993), and that governments marshalling material and discursive powers may be rewarded with improved levels of popular legitimacy.

Pelling and Dill (2009) approach disasters as both political events in and of themselves, and potential producers of secondary political effects (e.g. new alli-ances, leadership and social critiques) (ibid). Disasters triggered by environmental phenomena do not cause political change; rather they act as catalysts that put into motion potentially provocative social processes at multiple social levels.

Change reveals itself in policy discourse as well as in material or resource flows. Recent scholarly debate on this question has been influenced by two claims. The first sees disasters producing an "accelerated status quo"—here, change is path dependent on and limited to a concentration or acceleration of pre-disaster trajecto-ries, which remain under the control of powerful elites (Klein, 2007). The second viewpoint sees evidence that disasters can catalyse a critical juncture—an irrevers-ible change in the direction or composition of a political regime (or its subsets) (Pelling & Dill, 2009). Hence, crisis and disasters might act as assemblage convert-ers, in particular well-placed component parts resulting in transformations across several interlinked assemblages. Thus, although a particular social formation appears strong, it is always dependent upon and embedded within other structures and processes that have greater capacities for change (Haarstad & Wanvik, 2016). As we see in this chapter, the relationship between industry, government and indig-enous communities appears to have strong historical roots, but still, external and internal shocks of finance and fire are shaking the foundations of this relationship.

8.2   Immediate Consequences

Alberta had a mild, dry winter and spring, likely attributed to El Niño, where warmer than normal temperatures over the Pacific Ocean influenced weather all over North America. Probably caused by human carelessness, the wildfire wrecking huge areas in and around Fort McMurray had far-reaching effects. 2400 homes were lost, and the direct and indirect costs skyrocketed, beating all former Canadian natural disas-ters with its 9.9 billion Canadian dollars (Weber, 2017) (Fig. 8.1).

Five hundred First Nation and Métis firefighters worked on the Fort McMurray wildfire during the efforts to extinguishing the flames and limit their destructive impact. The Indigenous crews made up half of all fire brigades provided for battling the wildfire and comprised the largest percentage of firefighters working on the blaze. Indigenous firefighting crews were called "the backbone" of the forest fire

Fig. 8.1 The Beast (brown coloured area—Fort McMurray area hit by wildfire, main image from Landsat 8 false colour mosaic June 20. 2016, subset image from Sentinel 2 false colour mosaic May 3, 2016). Map by Benjamin Robson

service in Alberta (Pruden, 2016), and Indigenous communities and groups scrambled to ensure their people were safe. The communities also took in thousands of evacuees from Fort McMurray, as tens of thousands fled the massive wildfire raging around the city (Fontaine, 2016).

8.3   The Aftermath of Disaster

During an evening council meeting July 19, 2016, RMWB administration presented recommendations as to whether construction of approximately ninety capital projects should resume, be deferred or outright cancelled (Bird, 2016a). Halfway through the debate, one of the councillors made an amending motion to defer an additional ten projects which were previously scheduled to resume work. "Recovery is more than money. It's much more than money. We will need as much money as we can keep our hands on for that," the councillor claimed (Bird, 2016a). There was ample need for drastic budget cuts, as the councillor explained in an interview: "The city budget is moving towards a substantial deficit as an effect of the financial downturn, with huge expenditures and a strangled income revenue stream. In the current financial climate, we have to cut. We started out before the fire, cutting flat on all our administration around 10%" (Urban councillor, interview 2016). This decision effectively worked as a setback of previous, hard won gains for Indigenous communities.

The single most catalysing deferral of a proposed capital expenditure was that of a multi-million-dollar (Canadian) activity facility named the Conklin Multiplex. After significant rallying by Indigenous communities and local councillors for almost six years, a multi-use facility for the people of the predominantly Métis hamlet of Conklin had been negotiated over more than fifteen times in the Regional Municipal Council (Conklin Resource Development Advisory Committee (CRDAC), interview 2016), and the construction work had finally begun. The project deferral was to become the trigger of a large mobilization among indigenous communities across Wood Buffalo.

Three weeks after the council decision to defer the rural development projects, an opinion piece in *Connect Weekly*, a local paper and webpage, ignited the communities. It stated,

> The people of the community of Conklin were dealt a major blow on July 19 when RMWB Council voted to defer the completion of the Conklin Multiplex, a project that they have promised the community for over six years – an integral service hub for the community and industry. The hamlet of Conklin had high hopes as this project was finally going to provide a safe place for kids to play, a promising location to house nursing station and policing services, and essential programming for our youth and Elders (O'Donnell, 2016).

Conklin Resource Development Advisory Committee (CRDAC) is a multi-stakeholder entity put up to facilitate the development of the hamlet of Conklin in the wake of industrial developments. The CRDAC manager claimed that often, the voice of remote communities is small and goes unnoticed, and this has once again proven to be the case for the community of Conklin (interview 2016).

After the massive indigenous mobilization against the wildfire and, in Conklin and other outlying communities, to support the fleeing urban population from Fort McMurray, the deferral was particularly discouraging. "We stepped up and opened our homes and our wallets for Fort McMurray," said the president of Fort McKay Métis. The multiplex was the catalyst, concurred the CEO of the same community (interview 2016), and the communities decided that enough was enough (McDermott, 2016). Several Métis communities and other First Nations mobilized massively for the upcoming council meeting in downtown Fort McMurray. The council chambers' public seating area was packed with people as community leaders took turns speaking to council (Bird, 2016b).

The group included the Athabasca Tribal Council, the community associations for Conklin and Anzac, and all of Wood Buffalo's Métis locals. It became clear in the first sentences of McKay Métis president Ron Quintal's presentation that council had awoken a sleeping giant. The group turned the multiplex from a small issue of poor planning, into a massive issue of the overall negligence of rural areas that has persisted since 1995 (King, 2016). One of the rural councillors explains what happened:

When it came around to the next meeting, one of the councillors' (the one that broke the defence line of Conklin the last time) chief came along, to the meeting. He's First Nation, sitting right in the front row. He made sure that the councillor in question voted in favour of Conklin. And then there is this other councillor who flip-flops all the time, and he flipped to our side, and so then it was 7-3 in favour (of Conklin) (Rural councillor, interview 2016).

Although the councillor reported of a complex backstory of the first deferral, the massive presence of rural indigenous communities, and the particular effect of the First Nation chief being there turned the votes and secured the Conklin Multiplex. Once it became clear the Multiplex would return, delegates and rural residents gave a standing ovation. "It's a huge win, not just for Conklin but for the rural communities," Quintal said. The next step, he added, was to continue pushing the Minister of Municipal Affairs, Danielle Larivee, for a review of the entire amalgamation agreement of 1995, thereby calling the very existence and legitimacy of the RMWB, plus the regional primacy of Fort McMurray, into question.

With other grievances ranging from a lack of water and sewage services to the polarizing Conklin Multiplex, Wood Buffalo's rural communities petitioned Alberta Municipal Affairs to do a review of the amalgamation process back in 1995 to determine if they have been treated fairly under the terms of the negotiations. Representatives from nearly 20 First Nation and Metis groups, as well as the rural hamlets themselves, met with Larivee, arguing they are increasingly at an economic and social disadvantage compared to Fort McMurray (McDermott, 2016). In recent years, rural community leaders had forced quality of life issues into municipal political discourse. Access to piped sewage and running water has been a decades-long fight for some communities, and fresh food is still hours away for others.

McKay Métis president Quintal spoke of solidarity when he stated "We have been brought together by the fact that Conklin is a community in crisis and their needs are being ignored". He followed up by pointing to the cumulative effects of the grievances faced by the rural communities: "The problems they face have been amplified by the fire and the deferment of the multiplex has damaged the relationship that our communities have with the regional municipality" (Heidenreich, 2016). The timing is opportune. There was concern that a political focus on rebuilding Fort McMurray would further direct resources away from rural areas. Yet, Quintal said there was an opportunity to use the rebuild to reset the relationship between the urban and rural areas. "At the heart of the issue is the revenue provided by industry sites near our communities to the municipality and what we get in return," said Quintal. "At the end of the day, the rural communities are impacted by oil sands development and if the oil sands development are what's paying the taxes—the 95% of the taxes to the Regional Municipality of Wood Buffalo—why shouldn't the rural communities be able to prosper from that?" he said in an interview with a local newspaper (Heidenreich, 2016). By these quotes, it is evident that the consequences of the wildfire had reached its peak. From being a devastating natural disaster, it had turned into a powerful catalyst of change, pushing the rural, indigenous communities to opt for a fairer distribution of benefits and tax revenues within the entire regional municipality.

8.4   Analysing Politics of Disaster

Disaster politics analysis suggests that disaster events and their management are part of unfolding political histories (Pelling & Dill, 2009). The histories of the contested energy spaces of Wood Buffalo and the power relations between state, corporations and Indigenous communities clearly influence the impact and political aftermath of the wildfire disaster of 2016. Métis communities of Wood Buffalo have spent decades trying to improve their status as rights-bearing communities among the proliferating industrial developments in the Regional Municipality (Gaudry, 2014; Wanvik & Caine, 2017). There is little doubt that indigenous advocacy has had a substantive impact on the lives of local Métis communities all over Alberta, and in the way multinational companies and provincial and local governments deal with Métis issues. So, when the economic downturn was beginning to have a real impact on municipal revenues, followed by a devastating wildfire, political concessions were taken away or put on hold. Rural Métis and First Nation communities responded by rising to the occasion and fighting for hard-won access to the benefits of development. In a disaster politics framing, the council decision could be considered an attempt to accelerate the status quo, territorializing the energy assemblage to a previous state of affairs. The indigenous mobilization of resources creates destabilizing processes threatening to break with the status quo, first by rescuing the Conklin Multiplex from deferral, and subsequently by attempting to renegotiate the amalgamation agreement from 1995. The renegotiation is ongoing, and its outcome is still not known.

After years of feeling like an afterthought, leaders in the rural communities surrounding Fort McMurray say their relationship with the city is improving after facing the May 2016 wildfires (Beamish, 2018). Fort McKay took in approximately 5000 people during the evacuation. Fort McKay Metis president Ron Quintal said the event helped strengthen the whole region. "From a governance perspective, I think there's definitely been an improvement over the last couple of years," he said. "I think there's a genuine sincerity in terms of relationship. All the firefighters for the region have gotten that much closer with our Fort McMurray firefighters," said Quintal, who was also deputy chief of Fort McKay's department on the day of the fire. "The fact that we fought fire hand-in-hand during the Beast strengthened that bond, not just with Fort McMurray, but with all the other communities within the regional fire service" (President Quintal, quoted in Beamish, 2018). These events are all evidence of a particular mobilization of scarce, but at the same time potentially effective, human and material resources that enable marginalized communities to join forces and fight for their rights and title as Indigenous Peoples.

8.5   Conclusion

The activities of extractive industry have a great impact on the social, cultural and environmental realities in the carbonscapes of Wood Buffalo. Although there have been immense benefits for Canadian society, the burden shouldered by local ecosystems and indigenous communities is substantial, which adds to the prolonged historical conflict between the Crown and Indigenous Peoples in Wood Buffalo over rights and entitlements (Cairns, 2000; Veltmeyer & Bowles, 2014). Nevertheless, empirical evidence suggests that such a one-dimensional view on power dynamics fails to properly understand the encounters between industrial developers including the state and indigenous communities. In this chapter, I explored the case of Métis communities and their First Nation allies in the Regional Municipality of Wood Buffalo, specifically their mobilizing of resources in the wake of the devastating Fort McMurray wildfire of 2016. Here, local indigenous communities pooled their resources in order to exploit the events following the fire, in their struggle for survival and recognition.

Over the years, marginalized Indigenous peoples and their territories have forged a powerful assemblage in the contested energy spaces of Wood Buffalo. Through documentation of their connection with the land, local Metis communities have been exploiting regulatory windows of opportunity to opt for increased self-determination and empowerment (Keeling & Sandlos, 2016; Wanvik, 2016). However, as I have shown, the outcomes of these efforts have proven vulnerable to external shocks, like the economic downturn from 2014 and recently by the devastating Fort McMurray wildfire of 2016. By mobilizing scarce resources and responding to political territorializing efforts by the regional municipal council, local Métis groups joined forces and scrambled for increased political leverage: first by reversing the deferral of a local development project, the Conklin multiplex, and second by demanding renegotiation of the regional amalgamation agreement of 1995.

References

Alary, B. (2016). *Fort McMurray blaze among most 'extreme' of wildfires: researcher*. [News pages University of Alberta]. University of Alberta. Accessed 08 Aug 2016.

Albala-Bertrand, J.-M. (1993). *Political economy of large natural disasters: With special reference to developing countries, OUP catalogue*. New York, NY: Oxford University Press.

Alberta Energy. (2017). Resource revenues collected. Government of Alberta, Accessed October 20. http://www.energy.alberta.ca/About_Us/2564.asp.

Arendt, H. (1986). The totalitarian leader. Political leadership: A source book. Pittsburgh: University of Pittsburgh Press

Beamish, L. (2018, May 3). Relationship between McMurray, rural areas continues growing since wildfires. *Fort McMurray Today*. Retrieved from http://www.fortmcmurraytoday.com/2018/05/03/relationship-between-mcmurray-rural-areas-continues-growing-since-wildfires

Bebbington, A. (2012). Underground political ecologies: The second annual lecture of the Cultural and Political Ecology Specialty Group of the Association of American Geographers. *Geoforum, 43*(6), 1152–1162.

Bebbington, A., Hinojosa, L., Bebbington, D. H., Burneo, M. L., & Warnaars, X. (2008). Contention and ambiguity: Mining and the possibilities of development. *Development and Change, 39*(6), 887–914.

Bird, C. (2016a, July 20). Council defers Conklin Multiplex construction, other projects. *Fort McMurray Today*. Retrieved from http://www.fortmcmurraytoday.com/2016/07/20/council-defers-conklin-multiplex-construction-other-projects

Bird, C. (2016b, August 16). Rural coalition claims a victory as Conklin Multiplex Construction resumes, News report. *Fort McMurray today*. Retrieved from http://www.fortmcmurraytoday.com/2016/08/16/rural-coalition-wins-first-victory-at-council

Birkmann, J. (2006). *Measuring vulnerability to natural hazards: Towards disaster resilient societies*. Tokyo, Japan: United Nations University Press.

Black, T., D'Arcy, S., & Weis, T. (2014). *A line in the tar sands: Struggles for environmental justice*. Toronto, Canada: PM Press.

Cairns, A. C. (2000). *Citizen plus - Aboriginal peoples and the Canadian state*. Vancouver, Canada: UBC Press.

Connacher Inc. (2010). EIA: Appendix 7 - Traditional ecological knowledge and land use (TEK/TLU). Connacher Inc.

DeLanda, M. (2016). *Assemblage theory*. Edinburgh, Scotland: Edinburgh University Press.

Dembicki, G. (2012). Oil Sands Carbon: When Money and Morals Align. The Tyee. Accessed 20 Oct 2018. https://thetyee.ca/News/2012/06/19/Oil-Sands-Money-and-Morals/.

Eyford, D. R. (2015). *A New Direction - Advancing Aboriginal and Treaty Rights*. Canada: Aboriginal Affairs and Northern Development Canada.

Gaudry, A. J. P. (2014). *Kaa-tipeyimishoyaahk - 'We are those who own ourselves': A political history of Métis self-determination in the north-west, 1830-1870* [PhD Monograph]. Victoria, Canada: University of Victoria.

Haarstad, H., & Fløysand, A. (2007). Globalization and the power of rescaled narratives: A case of opposition to mining in Tambogrande, Peru. *Political Geography, 26*(3), 289–308.

Haarstad, H., & Wanvik, T. I. (2016). Carbonscapes and beyond conceptualizing the instability of oil landscapes. *Progress in Human Geography, 41*, 432. https://doi.org/10.1177/0309132516648007

Heidenreich, P. (Producer). (2016, August 16). *Rural and indigenous groups near Fort McMurray call for major changes to RMWB, cite unfair tax structure. [Video interview]*. Retrieved from http://globalnews.ca/news/2884793/rural-and-indigenous-groups-near-fort-mcmurray-call-for-major-changes-to-rmwb-cite-unfair-tax-structure/

Hewitt, K. (2014). *Regions of risk: A geographical introduction to disasters* (2nd ed.). New York, NY: Routledge.

Huseman, J., & Short, D. (2012). 'A slow industrial genocide': Tar sands and the indigenous peoples of northern Alberta. *The International Journal of Human Rights, 16*(1), 216–237.

Keeling, A., & Sandlos, J. (2016). Introduction: Critical perspectives on extractive industries in Northern Canada. *The Extractive Industries and Society, 3*(2), 265–268.

King, T. (2016, August 21). A welcome flip-flop, opinion piece. *Fort McMurray Today*. Retrieved from http://www.fortmcmurraytoday.com/2016/08/21/a-welcome-flip-flop

Keeping, J. (1999). The legal and constitutional basis for benefits agreements: A summary. *Northern Perspectives, 25* (4).

Klein, N. (2007). *The shock doctrine: The rise of disaster capitalism*. New York, NY: Macmillan.

Le Billon, P., & Carter, A. (2012). Securing Alberta's tar sands: Resistance and criminalization on a new energy frontier. In *Natural resources and social conflict: Towards critical environmental security* (pp. 170–192). Basingstoke, UK: Palgrave Macmillan.

LeClerc, E., & Keeling, A. (2015). From cutlines to traplines: Post-industrial land use at the Pine Point mine. *The Extractive Industries and Society, 2*(1), 7–18. https://doi.org/10.1016/j.exis.2014.09.001

Madden, J. (2016). Daniels v. Canada: A case of simple answers with significant consequences. Candian Lawyer, April 18. 2016.

McDermott, V. (2016, August 4). Rural hamlets asks province to examine amalgamation, News article. *Fort McMurray Today*. Retrieved from http://www.fortmcmurraytoday.com/2016/08/04/rural-hamlets-asks-province-to-examine-amalgamation

O'Donnell, J. (2016, August 5). Major blow to Conklin Multiplex, Public letter. *Connect Weekly*. Retrieved from http://www.fortmacconnect.ca/2016/07/major-blow-to-conklin-multiplex/

Özerdem, A., & Jacoby, T. (2006). *Disaster management and civil society: Earthquake relief in Japan, Turkey and India* (Vol. 1). London, UK: IB Tauris.

Pelling, M., & Dill, K. (2006). 'Natural' disasters as catalysts of political action. *Media Development, 53*(4), 7.

Pelling, M., & Dill, K. (2009). Disaster politics: Tipping points for change in the adaptation of sociopolitical regimes. *Progress in Human Geography, 34*, 21.

Pruden, J. G. (2016, May 20). Indigenous firefighters battling Fort McMurray blaze follow long Alberta tradition, News article. *The Globe and Mail*. Retrieved from http://www.theglobeandmail.com/news/national/indigenous-firefighters-battling-fort-mcmurray-blaze-follow-long-alberta-tradition/article30110469/

Sandlos, J., & Keeling, A. (2013). Living with Zombie Mines. Seeing the Woods: A Blog. https://seeingthewoods.org/2013/03/06/living-with-zombie-mines/.

Statoil Canada. (2013). *Statoil Community Report 2013*. Norway: Oslo.

Veltmeyer, H., & Bowles, P. (2014). Extractivist resistance: The case of the Enbridge oil pipeline project in Northern British Columbia. *The Extractive Industries and Society, 1*(1), 59–68. https://doi.org/10.1016/j.exis.2014.02.002

Wanvik, T. I. (2014). Encountering a multidimensional assemblage: The case of Norwegian corporate social responsibility activities in Indonesia. *Norsk Geografisk Tidsskrift - Norwegian Journal of Geography, 68*(5), 282–290. https://doi.org/10.1080/00291951.2014.964761

Wanvik, T. I. (2016). Governance transformed into CSR – new governance innovations in the Canadian oil sands. *The Extractive Industries and Society, 3*, 517.

Wanvik, T. I., & Caine, K. (2017). Understanding indigenous strategic pragmatism: Métis engagement with extractive industry developments in the Canadian North. *The Extractive Industries and Society, 4*(3), 595–605. https://doi.org/10.1016/j.exis.2017.04.002

Watts, M. (1983). On the poverty of theory: Natural hazards research in context. In *Interpretation of calamity: From the viewpoint of human ecology* (pp. 231–262). Boston, MA: Allen & Unwinn.

Weber, B. (2017, January 17). Fort McMurray wildfire financial impact reaches almost $10B, study says. *The Star*. Retrieved from https://www.thestar.com/business/2017/01/17/fort-mcmurray-wildfire-financial-impact-reaches-almost-10b-study-says.html

Chapter 9
Conclusions

Abstract Why do some indigenous communities support extractive industry developments on their traditional territories, despite substantial destruction of the local environment and traditional indigenous land use practices? This concluding chapter provides an overview of the arguments found in this book.

Keywords Energy · Extraction · Canada · Indigenous · Instability · Governance

Why do some indigenous communities support extractive industry developments on their traditional territories, despite substantial destruction of the local environment and traditional indigenous land use practices?

As I have indicated in the introduction of this book, I identify three sub-questions that inform my analysis of the main research question. These are: (1) How can the *socio-material complexity* of contested energy spaces be *conceptualized*? (2) How can *instability* and *potential for change* in contested energy spaces be identified? (3) How can the *power play* between industry, state and indigenous communities in the contested energy spaces of the Canadian North be *understood*?

Initially I discuss how to *conceptualize* the *socio-material complexity* of contested energy spaces. I employ assemblage theory to identify contested energy spaces as complex places or situations. I argue that to analyse and understand these complex situations, we need to equip assemblage theory with acknowledged geographical concepts of place (and materiality), scale (and networks) and power (as the mobilization of resources). I provide some analytical categories and tools to assist geographers in understanding contested energy spaces specifically, and we hope to contribute to the ongoing scholarly discourse of place.

Furthermore, I investigate how to identify *instability* and *potential for change* in contested energy spaces. Expanding on my initial reflections on contested energy spaces, I elaborate on their instabilities, underscoring that instead of techno-institutional complexes, regimes or a coherent systemic "fossil capitalism" held

© The Author(s) 2018, under exclusive licence to Springer International
Publishing AG, part of Springer Nature 2019
T. I. Wanvik, *Contested Energy Spaces*, SpringerBriefs in Geography,
https://doi.org/10.1007/978-3-030-02396-6_9

together by a co-articulation of institutions, infrastructures and practices (Huber, 2013; Unruh, 2000; Urry, 2013), carbonscapes are looser association of social and material elements drawn together and pulled apart by a range of forces. I argue that this is liberating because it frees us from the assumption that changes need to impact the fundamentals of larger socio-technical regimes to be significant. For me, the important point is to illustrate that contested energy spaces are fragmented, contested and converted at particular sites. Therefore, in contrast to Brenner, Madden, and Wachsmuth (2011), who suggest that assemblage thinking blunts critical sensibilities, I find that assemblage thinking helps to open spaces for negotiation and contestation. I argue that there is a normative rationale for shifting researchers' attention towards instabilities and change. Destabilizing the permanence of contested energy spaces may be productive in its own right. The emphasis on structural constraints runs the risk of reproducing the oil industry's carefully scripted narrative of its own inevitability. It is critical that the specific lens that spatiality affords geographers also should be used to identify the cracks in the wall and the leverage points for transformation.

Further, I debate how *the power play* between industry, state and indigenous communities should be *understood* in the contested energy spaces of the Canadian North but from two perspectives, or on two scales. On a macro-scale, I find that industrial activities have had great impacts on the social, cultural and environmental realities of the contested energy spaces. The burden has been substantial for local communities and has added to the prolonged historical conflict between the Crown and indigenous communities over rights and entitlements. This complex relationship has led to substantial challenges for all stakeholders. In response to these challenges, the federal duty to consult, along with provincial responsibility for environmental impact assessments and locally negotiated impact benefits agreements, has all been delegated to industry, where corporate social responsibility (CSR) and stakeholder management are an important centrepiece. This delegation has been legitimized on pragmatic grounds, underscoring the better position of industry to consult the indigenous communities, to assess its own impact and to negotiate compensation and benefits agreements. I have identified an interrelated, nested and multiscalar governance structure emerging from these four distinct governance features (consultations, EIAs, IBAs and CSR) that can be viewed as a joint mobilization effort by government, extractive industry proponents and indigenous communities to realize a workable, win–win regulatory environment in the contested energy space of Wood Buffalo.

On a micro-scale, the indigenous communities calibrate their own participation in the emerging governance processes in the contested energy space of Wood Buffalo to strengthen their negotiating power. I employ assemblage theory as a basis for an analytical framework to examine indigenous Métis communities in Wood Buffalo. I reveal that indigenous engagement with extractive industry developments is neither static nor responsive in character. Rather, indigenous communities creatively and proactively engage with extractive industry developments on their traditional territories as strategic pragmatists. Viewing the interactions between the component parts of the Wood Buffalo carbonscape as the workings of an unstable and changeable assemblage reconfigures the way in which we interpret indigenous engagement; we no longer see them as passive victims or as only responsive to external pressure. We now see indigenous communities as goal-motivated, pragmatic component parts of the Wood Buffalo carbonscape. Through strategic pragmatism, we show that indigenous communities have substantial transformative capabilities embedded in their traditional ways of life. I find that these capabilities have moved the Métis communities of Wood Buffalo into formalized alliances striving to evolve and change to harvest strategic resources for their own benefit.

The indigenous communities subject to this study favour high-impact industrial activities on their traditional territories for several specific reasons. First, the complexity exposed in contested energy spaces does not allow simplistic or conventional understandings of indigenous agency. Second, the governance innovations in Wood Buffalo entail different and non-traditional approaches by which different stakeholders seek benefits from a highly prosperous industrial adventure. Third, by underscoring the instability of contested energy spaces and their constituent parts, I have shown that indigenous communities are no less changeable or pragmatic than other stakeholders, striving to evolve and change to harvest strategic resources for their betterment.

9.1 Contributions

> The world is will to power, being is striving, and being is flows or becoming. (Deleuze, 2004)

Energy production is contested. We need to rethink and transform the ways in which we search for, produce and consume energy. To change, we need to scrutinize and understand our current contested energy spaces properly. Where are the weaknesses? Where are the strengths? Where are the cracks or the possibilities in the system that might permit transformation and change? In this study, I have the ambition of developing a theoretical and analytical framework for improving the understanding of barriers and challenges related to energy production and transformation. Contested energy spaces have proven to be an excellent starting point for such theoretical experimentation and production.

On the other hand, I have found geography to be particularly suitable as a cradle for new empirically grounded theoretical innovations around contested energy spaces. First, the theoretical toolbox of contemporary and historic geographical scholarship has proven to be of profound value for examining the processes of contested energy spaces. Second, the closeness of geography to empirical realities is vital when researchers bravely embark on journeys to construct new and useful theoretical frameworks. Third, the broad fan of methodological experience and knowledge that saturates geographical scholarship should enable all of us to produce relevant and intriguing data with which to analyse our common energy future.

By this study, I would like to have contributed to a body of research literature that explains and analyses processes of change in globalized places. I have sought to understand barriers and challenges related to governing contested energy spaces, and have drawn on experiences of what works. At the same time, I aimed to understand the challenges that society faces when approaching new energy sources to satisfy a limitless appetite for energy.

With my contribution underscoring the instabilities of contested energy spaces, and the complex assemblage of stakeholders and other component parts that constitute these spaces, I hope that in some way I have pointed to the changeability that lies underneath seemingly permanent and persistent power relations and energy fixes.

Additionally, I hope that my examination of the governance structures of the contested energy spaces of Alberta, Canada acts as a warning to all stakeholders about leaving governance processes to market forces alone. This can only increase the instability of these spaces. Instead, stakeholders should organize in multistakeholder or multisectoral forums, where negotiations between communities, the state and corporations should take place in a responsible, transparent and independent manner to secure benefits for all involved—stakeholders and other component parts. Unless proper participatory, consultative procedures are employed, and the government of Alberta develops a Métis consultation policy, all companies that hide behind the arbitrary processes developed by the Aboriginal Consultation Office regarding Métis consultation will risk having their project approvals quashed by the courts (Thompson, Shannon, & Hibbitt, 2017).

Finally, I hope to have opened a broader perspective on the discourse on subaltern agency, exemplified by my encounters with highly empowered, politically active indigenous Métis communities, and the workings of what I have termed *indigenous strategic pragmatism*. Indigenous communities should never be treated as subject to circumstance but as agents in their own right.

References

Brenner, N., Madden, D. J., & Wachsmuth, D. (2011). Assemblage urbanism and the challenges of critical urban theory. *City, 15*(2), 225–240.

Deleuze, G. (2004). *Anti-oedipus*. London, UK: A&C Black.

Huber, M. T. (2013). *Lifeblood: Oil, freedom, and the forces of capital*. Minneapolis, MN: University of Minnesota Press.

Thompson, C., Shannon, S., & Hibbitt, T. (2017). *Alberta Court of Queen's Bench confirms the scope of procedural fairness and evidence to trigger Alberta Crown's duty to consult in resource development applications*. Retrieved from http://blog.blg.com/energy/Pages/Post.aspx?PID=288

Unruh, G. (2000). Understanding carbon lock-in. *Energy Policy, 28*(12), 817–830.

Urry, J. (2013). *Societies beyond oil: Oil dregs and social futures*. London, UK: Zed Books.

Index